MATH AND SCIENCE ACROSS CULTURES

MATH AND SCIENCE ACROSS CULTURES

Activities and Investigations from the Exploratorium

Maurice Bazin, Modesto Tamez, and the Exploratorium Teacher Institute

THE NEW PRESS

expl◯ratorium

Published in the United States by The New Press, New York, 2002
Distributed by W. W. Norton & Company, Inc., New York

Page 176 constitutes an extension of this copyright page.

⚠ BE CAREFUL! The activities and projects in this work were designed with
safety and success in mind. But even the simplest activity or the most
common materials could be harmful when mishandled or misused. Use
common sense whenever you're exploring or experimenting.

To find out more about other Exploratorium educational resources,
visit us online at www.exploratorium.edu.

LIBRARY OF CONGRESS CATALOGING-IN-PUBLICATION DATA

Bazin, Maurice.
    Math and science across cultures: activities and investigations from the Exploratorium/
Maurice Bazin, Modesto Tamez, and the Exploratorium Teacher Institute.
        p. cm.
    Includes bibliographical references.
    ISBN 1–56584–541–2 (pbk.)
    1. Mathematics—Study and teaching. 2. Science—Study and teaching. I. Tamez, Modesto.
II. Exploratorium Teacher Institute (San Francisco, Calif.). III. Title.

QA11.2.B3653 2001
510'.7'1—dc21

                                                                                00–136455

This material is based on work supported by the National Science Foundation
under Grant No. ESI-9450279. Opinions expressed are those of the authors and
not necessarily those of the Foundation.

with additional support from
The U. S. Department of Education
The California Department of Education

The New Press was established in 1990 as a not-for-profit alternative
to the large, commercial publishing houses currently dominating the
book publishing industry. The New Press operates in the public
interest rather than for private gain, and is committed to publishing,
in innovative ways, works of educational, cultural, and community
value that are often deemed insufficiently profitable.

The New Press, 450 West 41st Street, 6th floor, New York, NY 10036

www.thenewpress.com

Printed in the United States of America

10 9 8 7 6 5 4 3 2 1

CONTENTS

Acknowledgments

The activities in this book were originally developed for classroom teachers—to help them integrate multicultural math and science into their curricula, to open a science-rich environment to all students, and to make science and math more relevant and approachable.

This book was made possible by the generous support of the National Science Foundation, and is the result of collaborative work with many teachers who have participated in Exploratorium Teacher Institute activities or have visited us and led workshops.

The teachers who helped test the activities developed for this project are: Rosa Aleman, Carl Araiza, Brenda Barnes, John Barnes, Mildred Bidal, Susan Boshoven, Jim Caddick, John Cook, Cesar Dayco, Elvira Espinoza, Sue Friedland, Rafael Gonzalez, Allen Graubard, Rosa Haberfeld, Sherry Hernandez-Woo, John Holly, Carolyn Jaramillo, Tim King, Melanie Kohler, Olivia Kuhnert, Jeanette Luini, Sylvia Mayo, Ann Noon, Sonia Norman, Vicente Oropeza, Malcolm Jim Perkins, Lisa Rizzo, Lori Rizzo, Andres Rodriguez, Cora Salumbides, Angel Sanchez, Sandy Siegel, Annie Stewart, Guillermo Trejo-Mejia, Teresa Trejo-Mejia, and Anna-Marie White. Vicente Oropeza, Paula Salemme, and Dan Sudran contributed from the Mission Science Workshop in San Francisco.

Our colleagues at the Exploratorium Teacher Institute also offered their full support in developing this material. They are: Chris DeLatour, Paul Doherty, Tien Huynh-Dinh, Karen Kalumuck, Lori Lambertson, Karen Mendelow Nelson, Eric Muller, Don Rathjen, and Linda Shore.

This book owes its current form to the efforts of Nancy Tune, who took on the task of turning three classroom-oriented prototype booklets into a book for a more general audience. At the Exploratorium, Bronwyn Bevan ably managed the development of the original materials that became the foundation for this book. Ruth Brown, with Ellen Klages and Pearl Tesler of the Editorial Department, edited and organized the final material; Diane Burk, Stephanie Syjuco, and Stacey Luce in the Graphics Department added design and illustration skills. Amy Snyder and Lily Rodriguez provided their photographic expertise. Kurt Feichtmeir oversaw the budget and coordinated efforts with The New Press, where Ellen Reeves and Barbara Chuang so ably guided the creation of this publication. Dennis Bartels, Rob Semper, and Goéry Delacôte provided the invaluable institutional support without which this book would never have become a reality.

About the Exploratorium

The Exploratorium is San Francisco's innovative museum of science, art, and human perception. Its more than 650 interactive exhibits are designed to attract, puzzle, challenge, and engage museum visitors as they investigate the diversity of the natural and physical world.

Physicist Frank Oppenheimer founded the museum in 1969 as a place to introduce people to science by encouraging self-discovery. Many of the first exhibits were based on Oppenheimer's own "Library of Experiments," which he developed during his years as a Colorado high school science teacher in the 1950s.

Oppenheimer believed that students learn best when they explore concepts and phenomena on their own. The qualities that make the Exploratorium a model for hands-on science—a respect for innovation and play, and a knack for finding new ways of looking at things—continue to reflect that belief.

But the Exploratorium is more than a museum. It is also a world-renowned center for science education. In 1984, the California Department of Education designated it as the state's first Regional Science Resource Center. Seven years later, under the leadership of its new executive director, Goéry Delacôte, the Exploratorium created the Center for Teaching and Learning, linking five internal programs that explore the intersection of the museum's informal learning environment with formal learning.

One of these programs, the Exploratorium Teacher Institute, provides ongoing opportunities for middle school and high school teachers to work with museum staff through summer, weekend, and after-school workshops. Teachers are engaged in a "learning-by-doing" approach to teaching general science, physics, life sciences, chemistry, and mathematics.

This book was created to help increase awareness of multicultural issues while making science and math more relevant and approachable to people of all backgrounds. The staff of the Teacher Institute worked with many teachers, locally and worldwide, to develop the culturally relevant activities presented here, with the hope that they will open a science-rich environment to all students.

How to Use This Book

At the Exploratorium, we believe that the best way to learn about something is to do it, not just read about it. That's why the activities in this book are hands-on, inquiry-based, and multicultural. You'll have the chance to make your own discoveries and create your own experiments. You'll find all the information you need in each chapter—including copies of some original ancient texts. All you'll need to get started is your own creative mind and a few simple materials you can find around the house or at the store.

At the beginning of each chapter is a world map, with stars highlighting a country or countries. In some cases—as in Chapter 5, about math in ancient Egypt—the star indicates that the whole chapter is about one country. In other chapters, such as Chapter 8, about weaving baskets, the activity may be relevant to many places and cultures throughout the world, so there are many stars on the map showing which countries are mentioned in the chapter. Take some time to look through the book and see which activities or places interest you the most, and start there.

In most activities, you'll find a number of these symbols: ◑ . Each one marks a good place to stop reading and try a suggested experiment or investigation. Other activities—such as making a Brazilian *cuica* (Chapter 2), or creating dyes from natural materials (Chapter 9)—take you through a hands-on discovery process from start to finish.

You may also occasionally see this safety symbol: ⚠ , which will alert you when extra precautions are advised.

Whether you want to find out more about your own cultural heritage or just discover some new and interesting ways of looking at the world, this book will be a wonderful journey to other places and times.

An Introduction for Educators

Why Use a Multicultural Approach in the Classroom?

All too often the science, math, and technology studied in schools is limited to the major successes of the Western world. Students are frequently faced with the challenge of learning in an environment that may undervalue or ignore their own cultural backgrounds. This book is designed to help you increase your awareness of multicultural issues for all students, while making science and math more relevant and approachable for children of color.

Throughout the history of humankind, people have tried to understand nature and use their understanding to make life more comfortable. If you think of science as our knowledge of nature, and technology as applications of that knowledge, you can quickly see that all human cultures have made use of science and technology. Cultures all around the globe and throughout time have been observing and exploring nature and developing technologies that help them in their everyday lives.

Claudia Zaslavsky, a New York secondary school math teacher and author, says that ethnomathematical perspectives are essential for a comprehensive curriculum. The author of *Africa Counts*, one of the first books to focus attention on other cultures' contributions to math and science, she says, "Students of many backgrounds can take pride in the achievements of their people, whereas the failure to include such contributions in the curriculum implies that they do not exist."

This book is intended to provide teachers with a resource to create student awareness of the contributions of many cultures.

Using an Inquiry Approach

Science and math teaching have made some evolutionary strides in the last few years. The terms "inquiry" and "hands-on" have become familiar to most teachers. Classrooms are becoming places for students to search for themselves, by themselves, asking questions of the world around them, working with real materials and objects. As a result, students and teachers are learning to work not only in a hands-on way, but also in a "minds-on" way, collaborating as learners and discoverers.

In our search for inquiry-based materials in the multicultural field, we found books that explained activities and concepts but did not give the reader an opportunity to explore them. Most of these books (and computer programs) presented lessons as successive pieces of information to be absorbed by the students. Some proposed exercises and posed questions about the text, but these fell short of allowing the student to become an active partner in discovery.

The activities presented in this book are both multicultural and inquiry-based; they offer discoveries to be made by the reader.

Suggestions on How to Approach These Activities

These activities, all of which have been tested with students in grades 4–12, are self-contained. You'll find all the information you need in each chapter. You'll be asked only to gather simple, inexpensive materials and, whenever appropriate, to make copies of the pages needed by the students.

When we made these activities available to teachers to test in their classrooms, many provided us with suggestions on how to approach them. We pass these suggestions on to you with the hope that they will aid in your own exploration.

Set the Stage

Many teachers found it was best to set the stage for these activities by pointing out to students that they were about to embark on a great adventure. This helped students and teachers realize that using this book requires a different kind of effort and concentration than other curricula.

In learning to read Mayan numbers or decode *quipu* knots, each class experiences many of the same difficulties and frustrations that the first "readers" of these stones or knotted ropes experienced. In the classroom, you can temper these difficulties by the enthusiasm you generate in your students.

Be Willing to Explore with Your Students

In leading these activities, it's important to realize that you don't need to have all the answers. It's hard to say "I don't know": When someone asks you a question, you naturally want to supply an answer. At the Exploratorium, we feel that saying "I don't know" is a necessary part of learning. And we feel that teachers—along with students— should be comfortable saying it. Of course, as a teacher, you'll want to follow up with "Let's see if we can find out together!"

Each of these activities has been tested many times at many levels, and each class has worked together to make its own discoveries and express them in original ways. The challenge for you as a teacher is to keep alive the ongoing discoveries taking place in each student's head—not by providing them with answers, but by giving them the tools and clues they need to keep learning and exploring.

Keep Track of What You Find

In the process of discovering and deciphering, it's important to take notes and organize data. You and your students will be looking at numbers and patterns that you may not be accustomed to seeing, so it's essential to write everything down. Later on, you and your team can try to make sense of it together.

Work in Teams

Self-discovery and inquiry can be exciting, but this method of learning does not always come easily, especially with new concepts and unfamiliar ways of organizing ideas.

Most scientific explorations are carried out in teams, where individual suggestions and viewpoints combine and build toward comprehension. Working in teams may help you and your class *slow down* and *discover*—or rediscover—the power of collaboration. It's important that each group shares its findings once the whole class finishes gathering data.

How These Activities Can Fit into Your Classroom

These multicultural lessons can contribute to a variety of different curricula, including social studies. Activities cover many aspects of mathematics, technology, and science. Chapter 2, "Cuica: Making Music in Brazil," for instance, introduces students to the science of sound; Chapter 9, "Dyeing: Colors from Nature," helps students work with some of the basic principles of chemistry; and Chapter 14, "Mud Bricks: Making African Houses Stronger," gives students an understanding of the interaction between culture and technology.

Many of the activities in this book emphasize the origins of numbers and counting, introducing the idea of symmetry and mathematical proofs, and reinforcing concepts of number bases and place value. Other activities help students practice a variety of skills, including measurement, estimation, and the manipulation of tools and materials.

During the discovery process, students will find themselves identifying patterns, gathering and interpreting data, and constructing and using lists and tables. They'll use logic to unravel puzzles and solve problems; they'll make and use many of the tools they'll need to gather information; and they'll compare their own experiences with the experiences of people in many different times and places. Most important, they'll gain a new appreciation for the ingenuity and resourcefulness of people all over the world.

Have Fun!

Introducing your students to the important understanding that all peoples have made scientific, mathematical, and technological discoveries allows students of all backgrounds to develop a broader understanding of and respect for peoples and cultures that are traditionally excluded from the science curriculum. You will allow your students to directly experience many of the same challenges, discoveries, and pleasures that other people have made in real-world settings. Have fun!

Patterns and Play

SONA

Sand Drawings from Africa

SONA Sand Drawings from Africa

In an area of southwestern Africa, there's a tradition of storytelling using sand drawings called *sona*, whose complex, interwoven patterns reveal some interesting mathematical ideas.

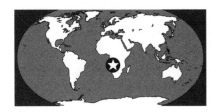

The Sand Drawings of the Chokwe

The Chokwe (pronounced "chock way") people live in southwestern Africa, mainly in northeast Angola. They are famous for their decorative arts, including beautiful woven mats and baskets, pottery and wood sculptures, and the striking geometric designs they use to decorate the walls of their homes.

The Chokwe also use their art in storytelling. They have an ancient tradition of making sand drawings, known in their language as *lusona* (plural: *sona*), to illustrate their stories. The sona illustrate proverbs, fables, games, riddles, and stories about animals.

In all cultures, there are stories that have been passed down through many generations, from the old to the young. Among the Chokwe, expert storytellers not only tell the tales well, but they also create traditional sona drawings to go with each story. These drawings have specific rules. The storyteller starts by making a series of dots, evenly spaced, in a rectangular array. Each drawing has a set number of rows and columns of dots.

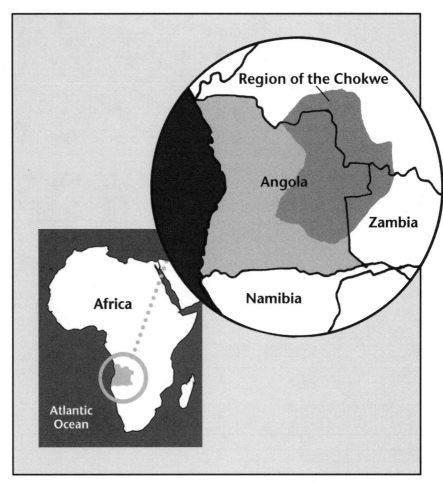

Once the narrator makes the dots, the drawing can begin.

The drawing itself is made up of lines that weave in and out around the dots. The narrator must draw and talk at the same time, without stopping. The listeners follow the storyteller closely, listening and watching as the drawing develops. If the storyteller hesitates, the audience smiles or laughs at the mistake.

In the past, the sona sand drawings and the stories they illustrated played an important role in passing knowledge from one generation of Chokwe people to the next. But several hundred years of colonization and slavery have weakened the tradition of making sand drawings. Many of the drawings are no longer made today. A lot of what we know about them comes from the records of missionaries, who wrote about the sona and copied some of the drawings on paper.

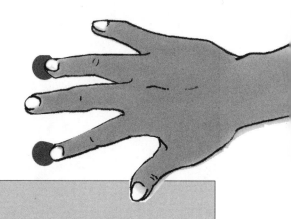

WHAT'S IT ALL ABOUT?

As you learn about the beautiful art of the Chokwe sona, you'll also explore the mathematical principles involved in creating these drawings. Along the way, you'll discover:

- How many unbroken lines it takes to make each sand drawing
- What algorithms are and what they have to do with sona sand drawings

Is there anything special I should know?

- This activity is recommended for ages 10 and up
- You can do this activity by yourself, or with friends

How much time will I need?

- This activity will take 1 to 2 hours

What materials will I need?

If you're working inside, you'll need:

- Graph paper
- Colored pencils

If you're working outside, you'll need:

- A place to make drawings in fine dirt or sand
- About 20 small stones, uncooked beans, or other small place markers

You'll also need:

- Copies of "The Stork and the Leopard" pattern, on page 9 (optional)

Making Sona Drawings

A simple sona drawing begins with dots arranged in a rectangular array. If you're using graph paper, mark the intersections of the lines on the paper with pencil dots. If you're outside, use the techniques that the Chokwe storyteller would use to make this array.

After cleaning and smoothing the sandy ground, the Chokwe storyteller first marks points with the tips of his extended fingers. He uses the index and ring fingers of his right hand to mark the points. His goal is to make an array of points that are evenly spaced, like the intersections on your graph paper.

To mark points from right to left, the storyteller keeps the tip of his ring finger on the last point he marked on the ground, and marks a new point with his index finger. He then moves his ring finger to the new point and marks another new point with his index finger. This method guarantees that the distance between points remains the same. When the storyteller marks points from left to right, he uses his ring finger to first mark the new points. To mark points going up or down, he simply turns his hand 90 degrees.

The storyteller now has an array of evenly spaced dots set up in a rectangular pattern. The size of the array depends on the story he is going to tell.

If you're outdoors, you can make the array with your fingers, then use small stones at each point to make the array easy to see. Try making your own array of three rows and three columns of dots.

Drawing a Sona Pattern

Once the points have been marked, the storyteller begins to draw, usually with the index finger of his right hand.

The drawings, which are a sort of language made up of points and lines, obey certain rules. They are always made with the fewest possible lines. Some drawings that may look very complicated can be made with only one line. The storyteller starts and finishes without ever having to lift his finger.

No matter how big the array is, or how complicated the pattern may look, both the storyteller and the Chokwe people watching know how many lines will be needed as soon as they see the array of dots. As you re-create some sona patterns, you'll begin to discover the secret of this understanding for yourself.

Before you try to draw a sona pattern of your own, you should know five rules:

1. DO NOT connect the dots! Sona lines go around and between the dots, but never touch them.

2. For most sona drawings, you can begin your first line between any two dots of the array.

3. Once you've picked a starting point, draw a straight line at 45 degrees between the dots. When you reach the edge of the array, turn your line 90 degrees and make another straight line.

4. You may cross a line you've already made, but don't trace over the same line twice.

5. If the end of a line meets its own beginning, that's a closed line. Once you've made a closed line, you may need to find another place in the array to begin another line. (If you're using colored pencils, use a different color for each closed line to help you see the patterns more easily.)

THE ANTELOPE

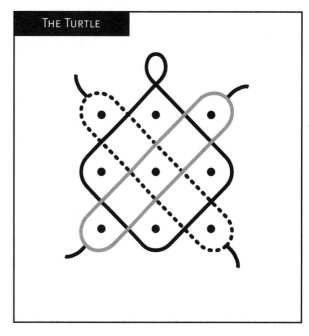

THE TURTLE

The Antelope

The sona pattern that depicts an antelope is drawn on a 3 x 4 array of dots. Make the array first and then make the drawing.

Remember that the lines go through the array diagonally. What is the smallest number of closed lines you can use to make the body of the antelope—not counting the head, legs, and tail?

You can actually make this drawing with one closed line. You can start anywhere on the body of the antelope and come back to that point without lifting your finger or pencil. When your line comes back to where you started, you've finished the body. Then you can add the head, tail, and feet to finish the antelope.

If your drawing required more than one closed line, go back and try it again. Remember that the drawing does not connect the dots. Instead it passes and loops around them. The lines end up making shapes with the dots in the center.

Notice that the lines you drew to make the animal are oriented at 45 degrees, diagonally to the array. If you're working with graph paper, you can use your pencil to draw a very faint rectangle around the outside of the finished drawing. Wherever you start drawing, the line should go straight until it reaches the edge of this invisible rectangle. When your line reaches an edge, it must make a 90-degree turn. At the corners, you have to make a U-turn—a 180-degree turn.

The Turtle

Take a careful look at the drawing of the turtle, above. You can see the points of a 3 x 3 array that the storyteller made before starting the drawing.

To draw the turtle, start by making your own 3 x 3 array of points.

The turtle is drawn on a smaller array than the antelope. How many closed lines do you think you'll need to draw it—not counting the four short legs?

You may be surprised to find that it takes three closed lines to draw the turtle. Each is shown as a different shade or pattern in the picture at the top of the page.

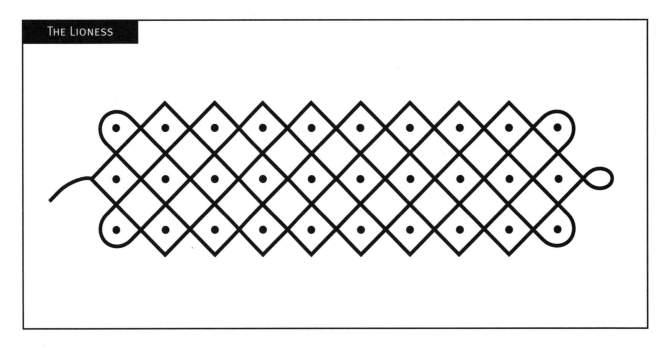

THE LIONESS

The Lioness

First make a 3 x 10 array. Then take a minute to study the picture of the lioness, above.

Now try re-creating the picture, using the same technique you used for the antelope. Start anywhere and draw a straight diagonal line, turning only when you reach a corner or an edge. ◑

You've seen that the lines of the sona are drawn at a 45-degree angle to the sides of the invisible rectangle. What other patterns or rules do you notice as you draw the lioness? ◑

The body of the lioness can be drawn with a single closed line. Just as in sona drawings of the turtle and the antelope, when a line reaches a side of the rectangle, it makes a 90-degree turn. When a line reaches a corner, it makes a rounded U-turn and heads back along a path parallel to the one it came in on. The head and tail are added at the end.

Sona Physics

There's a physics principle hiding in these drawings. Look carefully at the two drawings at the top of page 9.

They are the basic lioness and turtle but the corners are squared off and the invisible rectangle around each drawing is thinly marked.

All the bouncing lines may remind you of billiard balls moving on a pool table or light rays hitting a mirror: When they strike a surface at an angle of 45 degrees, they bounce off at the same angle. Notice that the lines in these drawings follow the same kind of trajectory, traveling straight until they reach the edge and then bouncing off at an angle equal to the incoming angle. Once we draw the invisible rectangles around the drawings, the set of rules we used initially become equivalent to the physical laws of reflection on a surface.

Rules Make Rhythm

The sequence of the movements that you make as you draw the lioness, the turtle, and the antelope follow rules or equivalent physical laws. These rules determine a *rhythm*, specific repetitions that are independent of the size of the rectangular array of points and of the number of lines necessary to com-

plete the drawing. The rhythm used to draw these three sona is different, for instance, from the one you'd need to use if you were drawing the Large Lion's Stomach shown on page 13.

In a sense, the rhythm generated by the repeating motions you make to complete the drawing is similar to the repeated use of a mathematical operation. This is called *executing an algorithm*. An algorithm is a set of operations (like "add two, then divide by ten") that gets you from some starting situation to a result. A key aspect of algorithms is that they always apply the same rules in any situation, but give different answers depending on the starting conditions. In our example, if you start with 18 and apply the algorithm "add two, then divide by ten," you get 2. If you start with a different number, you get a different result.

Although all these sona designs followed the same rhythm, the number of closed lines you needed differed, depending on the number of dots in the rows and columns of the array you started with.

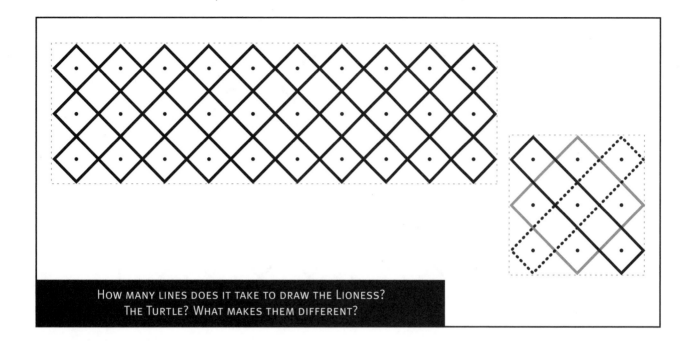

How Many Lines?

You may have been surprised to find that the lioness and the antelope could be drawn with only one closed line, while the smaller turtle required three. The number of closed lines you need to make each drawing depends on the dimensions of the array of points you start with. For example, you need only one closed line to draw a baby lion with a 3 x 5 array. Yet if you try drawing a larger lioness with a 3 x 9 array, you'll find that you need three different closed lines to complete your drawing.

In Angola, Ghana, Congo, and other African countries, many adults and children know how many closed lines are needed to make a drawing as soon as they see the array of dots. If you ask how many closed lines are needed to make a sona drawing with a 4 x 6 array, they can tell you right away that two lines are needed. Show them a 5 x 7 array, and they quickly say that one closed line will be enough.

The Stork and the Leopard

Once the leopard Kajama asked the stork Kumbi for some feathers to line his den. Some time later, the stork asked the leopard for a piece of his skin. When Kajama granted the stork's request, he died. Kajama's sons tried to take revenge, but Kumbi, who knew the swamp very well, was able to escape.

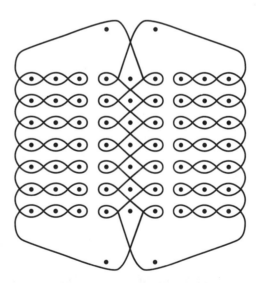

IN THIS DRAWING, THE WINDING LINE IS THE PATH OF THE FLEEING STORK, KUMBI. THE POINTS REPRESENT THE SWAMP THROUGH WHICH KUMBI MAKES HIS ESCAPE. THE DRAWING ACTUALLY CONSISTS OF TWO INTERTWINING CURVED LINES. YOU MAY WANT TO TRACE OVER THE ESCAPE ROUTE BY FOLLOWING IT WITH TWO DIFFERENT COLORED PENCILS ON A PHOTOCOPY OF THIS DRAWING.

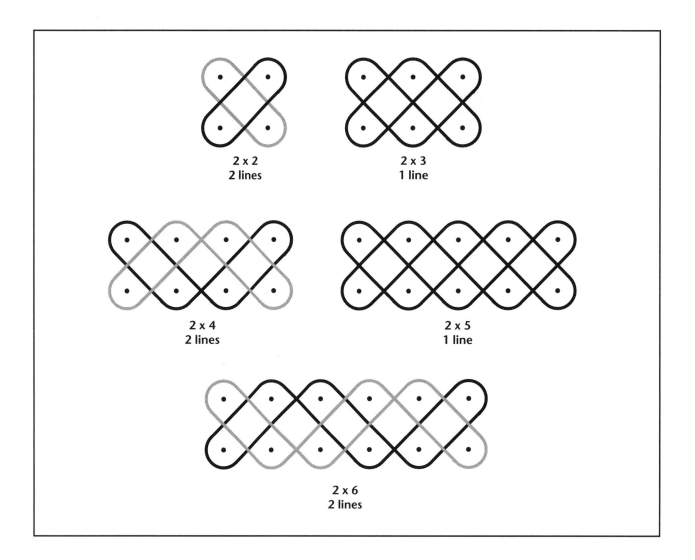

2 x 2
2 lines

2 x 3
1 line

2 x 4
2 lines

2 x 5
1 line

2 x 6
2 lines

How can you look at an array of dots and tell how many closed lines will be needed for the drawing? To explore this question, you'll have to do some experimenting. Start with arrays based on two rows. Make arrays of 2 x 2 dots, 2 x 3 dots, 2 x 4 dots, 2 x 5 dots, and 2 x 6 dots. Then follow the rules to make a drawing for each array. What do you notice? ◖

It turns out that arrays of two rows with an even number of dots in each row (two, four, six) need two closed lines. But when there's an odd number of dots in each row (three, five, seven), the drawings need only one closed line.

How many closed lines do you think you'll need for a 2 x 8 array?

How about a 2 x 7 array? Test your ideas by doing the drawings. ◖

You should discover that you need two closed lines for the 2 x 8 array and one for the 2 x 7 array.

Now take some time to explore arrays with three rows. What kind of pattern or patterns can you find for them? ◖

The pattern is different here. If you have three rows, you need three closed lines to draw the figure if there are three or six points in each row. But if there are four or five points in each row, only one closed line is needed. Now do some exploring with arrays that have four rows. ◖

The pattern is different once again. In the 4 x 3 and 4 x 5 arrays,

one closed line is enough. In 4 x 2 and 4 x 6 arrays, two lines are needed. In the 4 x 4 array, you need four closed lines.

It seems as if a different rule applies for each size of array. But that's not the case. There are many patterns, but one way to predict the number of closed lines needed to draw a pattern.

To solve the problem, think about what you already know. For each array, you start with two numbers—the number of rows and the number of columns. From these two numbers, you want to find a third number: the number of closed lines needed to do the drawing. How can you figure this out? ◖

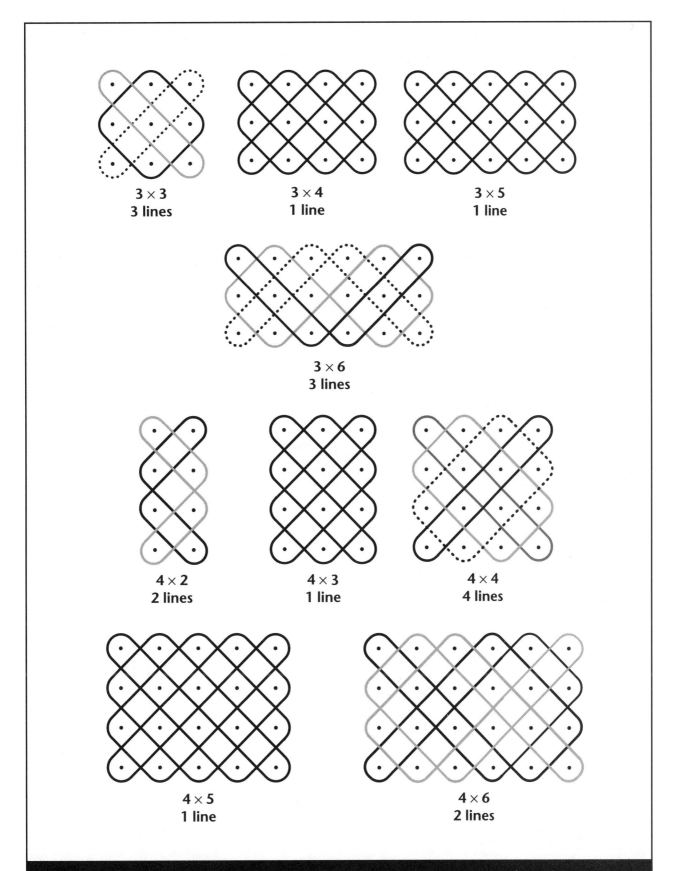

3 × 3
3 lines

3 × 4
1 line

3 × 5
1 line

3 × 6
3 lines

4 × 2
2 lines

4 × 3
1 line

4 × 4
4 lines

4 × 5
1 line

4 × 6
2 lines

HOW MANY CLOSED LINES DO YOU NEED TO COMPLETE EACH PATTERN?
IT DEPENDS ON THE NUMBER OF ROWS AND COLUMNS IN EACH ARRAY.

Number of Rows	Number of Columns	Number of Closed Lines
2	2	2
1	3	1

CAN YOU FIGURE OUT HOW THE NUMBER OF CLOSED LINES YOU NEED TO DRAW A LUSONA IS RELATED TO THE NUMBER OF ROWS AND COLUMNS IN THE ARRAY?

In mathematics, there are many ways to get one number from two others. For example, you could add the first two numbers or subtract one from the other; you could multiply or divide the numbers; you could find the lowest common multiple of the two. You can probably think of many other ways to get a number from two other numbers.

When you have several groups of three numbers and you're trying to figure out the rule that connects them, it's often helpful to make a table. Try making a table for the arrays you've worked with. Set up your table with three columns: Number of Rows, Number of Columns, and Number of Closed Lines, as shown above. 👋

Now see if you can find a mathematical relationship between the numbers in the third column of your table and the numbers in the first and second columns. You're looking for a relationship, or rule, that is true for every row in your table. Is the third number the sum of the two others? Is it the difference between them? Or is the rule more complicated?

If after a while you don't see a rule, don't give up. A lot of work in math and science involves exploring a set of data until you see a pattern that was there all along. Keep trying! Your work will be rewarded with the "Aha!" of discovery. 👋

Did you discover the pattern? Here it is: The third number is the largest number that divides evenly into the first two numbers. That is, the number of closed lines is equal to the greatest common divisor of the number of rows and the number of columns in the array.

For example, the greatest common divisor of 3 and 6 is 3, so it takes three closed lines to fill a 3 x 6 array. The greatest common divisor of 3 and 5 is 1, so you know you need just one closed line to complete a 3 x 5 drawing. What is the greatest common divisor of 4 and 2? Does that fit with the results on your table? 👋

The greatest common divisor of 4 and 2 is 2—and it takes two lines to complete a 2 x 4 array.

Look again at your table to be sure the rule holds true for all the arrays. You can test the rule by making other drawings with arrays you haven't tried yet. For example, how many lines should it take to draw a 3 x 12 lioness? Try it and see! 👋

FLYING BIRD

CHASED CHICKEN'S PATH

ANTELOPE FOOTPRINT

LARGE LION'S STOMACH

SPIDER IN ITS WEB

More Sand Drawings

These symmetrical Chokwe drawings are more complex than the rectangular arrays you have been working with. Each style—the Chased Chicken's Path, the Antelope Footprint, the Flying Bird, the Spider in Its Web, and the Large Lion's Stomach—has different rules of its own that must be followed by the storyteller. Enjoy these other styles of sona drawings for their intricate beauty, or explore further and see if you can re-create any of them.

This activity was developed by Paulus Gerdes.

Making Connections

• Does your family keep any photo albums? Are there family stories that go with the pictures? How did you learn those stories? Think about what might happen if one of your descendants looked at those albums a hundred years from now. Do you think they will know those stories, too? How?

• You probably learned many fairy tales when you were young. Ask people who come from different parts of the world about the stories they learned when they were young. Are they the same kinds of stories as yours?

• Many stories are told to help people learn a lesson or understand a problem. What have you learned from the stories, proverbs, and fables you have been told?

Recommended Resources

Frankenstein, Marilyn, and Arthur Powell, eds. *Ethnomathematics*. New York: State University of New York Press, 1997.

Gerdes, Paulus. *Geometry from Africa: Mathematical and Educational Explorations*. Washington, DC: Mathematical Association of America, 1999.

Jordan, Manuel. *Chokwe* (Heritage Library of African Peoples/Central Africa). New York: Rosen Publishing Group, 1997.

Jordan, Manuel et al. *Chokwe: Art and Initiation Among Chokwe and Related Peoples*. Prestel USA, 1998.

Zaslavsky, Claudia. *Africa Counts: Number and Pattern in African Culture*. New York: Lawrence Hill Books (3d edition), 1999.

2

CUICA

Making Music in Brazil

CUICA Making Music in Brazil

If you look at musical instruments around the world, you might be surprised at how similar they are. Though they make different kinds of music and are used in different traditions, they make sounds in the same ways.

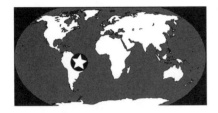

A BRAZILIAN CUICA

The Brazilian Cuica

One of the biggest celebrations in the world is the annual *Carnaval* in Brazil. During this four-day festival, which marks the beginning of the Catholic holiday of Lent, the streets of Brazilian cities are filled with people. In Rio de Janeiro, the highlight of Carnaval is an enormous parade lasting three days and nights. Groups of as many as a thousand people from each neighborhood get together to take part in the parade. The groups, called samba schools, wear elaborate costumes, dancing and playing music along the parade route. A basic part of each school's music section is an instrument called the *cuica* (pronounced "kwee-kah"). No samba school is complete without a cuica or two.

Like many other instruments, the cuica was brought to Brazil from Africa. Originally, it was made from a hollowed-out log with an animal skin stretched over one end and a rope attached to the center of the skin. In Africa, the instrument was used as a hunting tool. The sounds it made were like the howling of a female lion during mating season, so hunters used it to attract male lions.

The Brazilian cuica is a metal cylinder with a skin stretched over one end. Fixed at the center of the skin is a stick, usually made of bamboo. To play the instrument, the musician slides a piece of wet leather or cloth up and down the bamboo stick.

It's hard to place the cuica in any of the standard categories of musical instruments. It looks a lot like a drum, but you don't beat it, so it's not a percussion instrument. Like a clarinet or flute, it has an empty cavity in which air vibrates, but you don't blow air into it, so it's not a wind instrument. The stick inside isn't plucked or bowed like a string instrument; instead, it's rubbed to make the sound.

During a Carnaval parade, a musician wears a strap around his neck to support his cuica. The instrument rests on his chest, and he uses both hands to play it. One hand rubs the stick while the other hand holds the skin-covered end steady, pressing near the center of the skin to change the pitch.

Depending on how it is played, a cuica can produce an amazing variety of sounds. It can sound like a dog barking or a frog croaking. It can even imitate the sound of a person laughing or crying. The sound of a cuica can be rhythmic, almost like a percussion instrument, or it can be quite melodious.

THE PHOTOGRAPH ON THE RIGHT SHOWS HOW A CUICA IS HELD AND PLAYED. ABOVE IS AN INSIDE VIEW OF A CUICA. YOU CAN SEE THE STICK PIERCING THE CENTER OF THE SKIN.

WHAT'S IT ALL ABOUT?

In this activity, you will build and study a Brazilian instrument called the cuica. You'll investigate the science of sound and get a taste of the Brazilian celebration known as Carnaval. As you make and play your cuica, you'll have a chance to investigate:

- How a cuica produces sound
- How other instruments make sounds
- How you can change the sound a musical instrument makes

Is there anything special I should know?

- This activity is recommended for ages 10 and up
- You can work alone or with friends to create a cuica band

How much time will I need?

- Making your cuica will probably take about 30 minutes

What materials will I need?

- A large empty can with a plastic lid, like a coffee can (if you're working with a friend, use different sizes of cans to make different sounds; each can must be large enough to put your hand inside)
- A bamboo skewer about 8–10 inches long (like the ones sold in grocery stores for making shish kebabs); you may want to cut off the sharp end, or wrap it with tape
- A piece of damp cotton cloth, such as a piece of old towel or T-shirt
- White glue   • String or rubber bands
- Can opener   • Electrical or duct tape
- Small nails   • Hammer
- Guitar (optional)

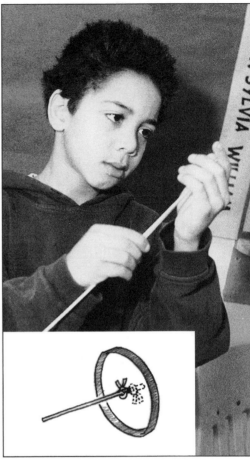

Building a Cuica

Follow these steps to make your cuica:

1. Take the plastic lid off the can and carefully put heavy tape over any sharp edges on the empty can. Then use the can opener to cut out the bottom of the can, so you have an open cylinder. Now carefully put heavy tape over the bottom edges, too. ⚠

2. Use a hammer and nail to make a small hole in the center of the plastic lid. Make the hole just big enough to push one of the bamboo skewers through, but not so big that the skewer will easily fall through.

3. Push the skewer through the lid, leaving a quarter-inch of the skewer sticking out the top of the lid.

4. Hold the skewer in place so it won't slip through the lid. Take two short pieces of string and tie two knots around the skewer, one directly above the lid and one directly below it. Tie each knot as close to the lid as you can. A rubber band doubled many times will also work.

5. To keep the knots from slipping, put a few drops of glue on each knot. Or you can carve two small notches in the skewer where the strings will be knotted.

6. Put the lid back on the can. If the skewer sticks out the bottom of the can, break off the extra piece.

7. The lid should fit tightly on the can. If it doesn't, put some glue along the sides of the lid to hold it on.

You have now completed your cuica.

Making Music

To play the cuica, hold the can with one hand and hold a moistened cotton cloth in the other. The hand with the cloth will be inside the can. Rub the cloth up and down on the long end of the bamboo skewer.

Cuicas can make sounds that range from deep croaks to high-pitched cackles. See what different kinds of sounds you can make. Experiment by holding the skewer tightly or loosely, or try rubbing faster or slower. What happens? ◖

If you and a friend both made cuicas, listen to the sounds the different instruments make. ◖

A Brazilian cuica player sometimes presses near the center of the skin with his free hand to change the sound. Try this by pressing gently near the center of the plastic lid while you play. What result do you get? ◖

The Science of Sound

To understand how your cuica works, you need to know a little about sound.

Sound is a vibration that travels. When you strike a drum, for example, the drumhead starts to vibrate. The vibrations push and pull on the surrounding air, causing the air to vibrate. The vibrations travel through the air and reach your ear, where they cause a thin membrane—your eardrum—to vibrate. If the rate of vibration is within a certain range—from 20 to 20,000 vibrations per second—then you hear a sound. The higher the rate of vibration, the higher the pitch of the sound you hear.

When you rub the bamboo skewer of your cuica with a wet cloth, the cloth sticks and slides rhythmically along the skewer. The rhythmic vibration caused by the sticking and sliding of the cloth creates sound. When you hear chalk screeching on a blackboard, you're also hearing sound made by sticking and sliding. The chalk sticks and slides on the blackboard at a rapid rate, so the sound produced is high and piercing. In the case of the cuica, the sticking and sliding of the cloth happens at a slower rate, so the sound has a lower pitch.

The vibrations in the cuica begin in the bamboo skewer, but the skewer isn't the only thing that vibrates. If it were, everyone's cuica would sound the same. The vibrations of the skewer make the plastic coffee-can lid vibrate as well. That changes the sound and amplifies it—that is, makes it louder.

The metal cylinder of the can creates an air space in which the sound can build. This hollow space is one of the most important parts of the instrument.

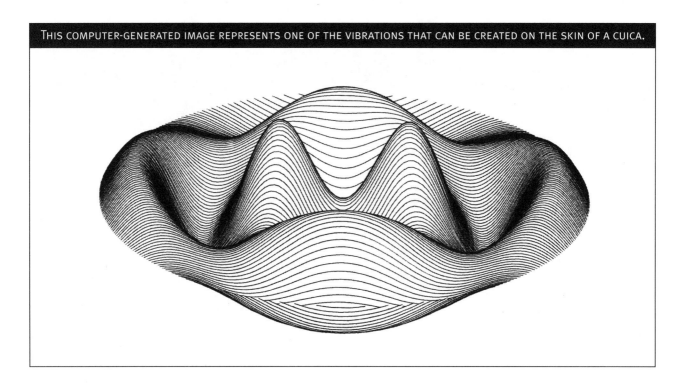

THIS COMPUTER-GENERATED IMAGE REPRESENTS ONE OF THE VIBRATIONS THAT CAN BE CREATED ON THE SKIN OF A CUICA.

In most instruments, there is a part that vibrates to make the sound and a large hollow space that amplifies the sound. In a saxophone, for example, a reed vibrates and makes the sounds that are amplified in the hollow, curved-tube body of the instrument. In a guitar, strings vibrate to make the sounds that are amplified inside the guitar's hollow body.

If you made more than one cuica, you may have noticed that larger cuicas make lower-pitched sounds, and smaller cuicas make higher-pitched sounds. In fact, this rule is true for all instruments: Large instruments make lower-pitched sounds than small ones.

The reason for this is the different sizes of the hollow spaces in the instruments. Inside the hollow space, traveling sound vibrations bounce back and forth, gaining strength. In a smaller space, the vibrations have less distance to travel, so they bounce back and forth more often. This results in a faster rate of vibration—and a higher-pitched sound.

Picture a cello and a violin. They are made in the same way and they have the same shape. The cello is really just a giant version of a violin. But they sound different from each other. Can you figure out why? ◗

Because its strings are lighter and its hollow body is smaller, a violin makes a higher-pitched sound than a cello. In the larger cello, vibrations bounce back and forth along the strings and inside the cavity at a slower rate, so the sound you hear has a lower pitch.

Advanced Cuica: Fundamentals and Harmonics

Did you notice that when you pressed on the coffee-can lid while playing the cuica, you made the pitch rise? If you have a guitar, you can use it to understand why this happened. Pluck any string about one-third of the way from either end and listen. The sound you hear isn't a pure note. Instead, it's made of many pitches. The lowest of these

pitches, called the *fundamental*, corresponds to the entire string wagging back and forth. The higher notes, called *harmonics*, correspond to smaller segments of the string wagging back and forth.

The fundamental and harmonic vibrations of the string happen at the same time (and quickly!), so it's not easy to see them. But you can hear them. If you pluck the string again and immediately touch the string lightly in the middle, you will hear the pitch rise. This happens because your finger makes it harder for the entire string to vibrate. But the smaller segments can still vibrate. So instead of hearing the fundamental, you hear only some of the higher-pitched harmonics.

The cuica is more complex than this. It vibrates all over the surface of the lid, not just along one line, as the guitar string does. Still, the basic idea is the same: When you press lightly at the center of the cuica lid, you stop the main vibration, the fundamental. The pitch you hear rises because only some of the higher harmonics remain.

This activity was developed by Maurice Bazin and Modesto Tamez.

Making Connections

• What instruments can you think of that are tied to particular celebrations or particular styles of music? What instruments would you be surprised to see in a rock band? In a marching band? When are you most likely to hear an organ play?

• Think about different musical instruments. What parts are vibrating? What materials could you use to duplicate those parts? Try experimenting with objects around you to make instruments of your own.

• Some sound artists make music from the sounds they hear around them, like the sound of a tennis ball thumping back and forth between two players, the thud of a ballerina's feet as she dances, the pattern of sound made by cars honking in a city. As you move from place to place, try closing your eyes and concentrating on the sounds around you. Can you hear any of this "ambient music"?

Recommended Resources

Darling, David J. *Sounds Interesting: The Science of Acoustics.* New York: Dillon Press, 1991.

Mattingly, Rick, ed. *Airto, the Spirit of Percussion.* Wayne, NJ: 21st Century Music Productions, Inc., 1985.

Nettl, Bruno, ed. *Excursions in World Music.* New York: Prentice Hall (2d edition), 1996.

Pierce, John R. *The Science of Musical Sound.* New York: W. H. Freeman, 1992.

3

MADAGASCAR
SOLITAIRE

Playing Games

MADAGASCAR SOLITAIRE
Playing Games

A simple game of jumping and removing pegs from a board can lead you to some interesting mathematical concepts.

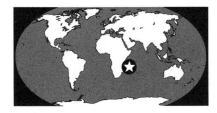

A Little About Madagascar

Madagascar is a very large island, almost a thousand miles long, located off the southeast coast of Africa. The island is famous for its unique plants and animals. In Madagascar, there are five varieties of baobab tree, for example, while in all the rest of Africa there is only one.

Until relatively recently, no humans lived on Madagascar. The first people arrived around two thousand years ago, most likely from Africa, India, and the Middle East. They found dense rainforests and an abundance of unique wildlife, including lemurs, dwarf hippos, giant tortoises, and ten-foot-tall elephant birds. At that time, there were probably more than one hundred species of animals found only on Madagascar. After a thousand years of human hunting, almost a quarter of those species had become extinct.

Today, the human culture of Madagascar is diverse, reflecting the origins of the Asian and African people who settled there over the last millennium. Almost everywhere

MADAGASCAR SOLITAIRE

in Africa you can see the influence of the European powers that colonized the continent. In Madagascar, the strongest European influence came from the French, who established a colony there in the late 1800s.

The game known as Madagascar Solitaire is a form of peg solitaire, a one-person game played by putting pegs in holes. In peg solitaire, the player tries to win against the game itself by moving and removing pegs (or other objects) from a playing board. In most of these games, when only one peg remains, the game is won.

Versions of peg solitaire are found in many parts of the world and come in a variety of forms. A modern version is played on a triangular board. No matter what the configuration of the board, the moves are simple, yet the game is always challenging.

Like many other games, Madagascar Solitaire demonstrates patterns and rules based on important mathematical concepts. Playing this game is useful for studying math because you can create as many versions as you want, varying the level of complexity as you master the simple principles involved.

Learning the Game

Step 1: Creating the Playing Area

To begin, mark your game's playing positions by drawing an array of small, equally spaced circles or dots on a piece of paper.

Your "game board" can have many different shapes and numbers of dots. Traditional Madagascar Solitaire is played on a 7 x 7 square array with the corners removed, like the one illustrated below.

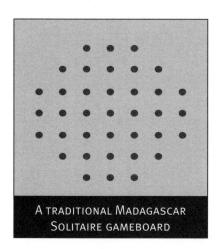

A TRADITIONAL MADAGASCAR SOLITAIRE GAMEBOARD

This is a pretty complicated board to start with, so we'll begin with a simpler array:

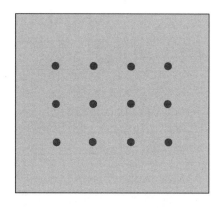

Step 2: Setting Up the Playing Pieces

Place a playing piece on every point on your array.

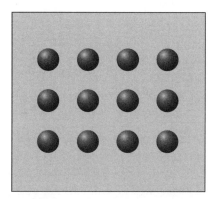

Step 3: Beginning the Game

Choose any one piece and remove it (**x** = playing piece; **o** = empty position).

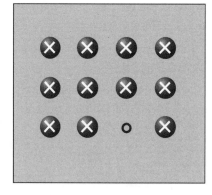

Step 4: Making Moves

For this version of the game, there are three simple rules:
- You can only move in a straight line.
- You can only jump over one piece at a time.
- You must land in an empty space.

To move, you simply jump one piece over another, then pick up the piece you jumped over, removing it from the game.

WHAT'S IT ALL ABOUT?

In this chapter, you'll learn how to play a game called Madagascar Solitaire. As you play, you'll explore concepts of game strategy and discover:

- How Madagascar Solitaire works
- How a simple version of a game can help you learn strategies for more complicated versions
- Whether or not you can know if it's possible to win a particular game—before you've even started

Is there anything special I should know?

- Although Madagascar Solitaire can be played by young children, this activity is recommended for ages 12 and up
- You can do this activity by yourself or in a small group

How much time will I need?

- You can play a game of Madagascar Solitaire in just a few minutes, or you can organize tournaments that might take longer

What materials will I need?

- Plain paper or graph paper
- Pencils
- About 40 small objects to use as playing pieces; coins, beans, shells, and pebbles work well

Step 5: Winning the Game

Each time you make a move, one fewer playing piece will remain on the board. To win the game, you must end up with only one piece on the board after you make your last possible jump.

Starting Simple

The simplest possible version of the game has a game board with just three points in a line, like this:

You may want to begin by playing a few games of this version. As you experiment, you'll quickly discover that if you start the game by removing either one of the end pieces, you MUST WIN. But if you start the game by removing the center piece, you MUST LOSE.

Now try the same game with four points in a row. 👌

In the 1 x 4 version, removing either end piece first means you MUST LOSE; removing one of the two inside pieces first means you MUST WIN.

Getting Off to a Good Start

In any version of peg solitaire, winning or losing depends upon the sequence of moves you make. The beginning points on a playing array fall into three categories:

1. Some points are MUST WIN points.

This means that if you start by removing a piece from one of these points, you'll win the game no matter what your next moves are. We'll use the letter "W" to label these points.

Recording Your Moves
Method 1:

Rows are labeled by letters, and columns are labeled by numbers; **o** marks the starting point.

For this system of labeling, a list of moves would look like this—the first label is the starting point, the second is the landing point:

A2–A4   C3–A3   B1–B3   etc.

Method 2:

Label both rows and columns with numbers; **o** marks the starting point. For this system of labeling, a list of moves would look like this:

1,2–1,4   3,3–1,3   2,1–2,3 etc.

As in any "ordered pair" situation, you have to be consistent about whether the number for the row or column comes first when you use this system.

Method 3:

The sites are labeled with a simple sequence of numbers. Here, the 4 is the starting point. For this system of labeling, a list of moves would look like this:

2–4   11–3   5–7   etc.

It's harder to see the pattern in the numbered moves, but it's there. For example, 2–4 and 5–7 are moves across columns in increments of two, and 11–3 is a move across rows, an increment of four times two.

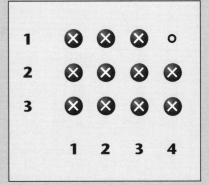

Tips from a Teachers' Workshop

At a 1993 staff development workshop at Nkrumah Teachers' Training College in Zanzibar Town, Tanzania, Madagascar Solitaire was used as an example of an accessible hands-on activity that tackled mathematical issues. Participants who played the game recorded their moves, the point they used to start the game, and the final positions of the pieces that were left when the game was over. Several methods were developed to do this; three are included here. These schemes may help you invent your own recording methods.

2. Some points are MUST LOSE points.

This means that if you start by removing a piece from one of these points, you will lose. No matter what moves you make, you'll have more than one piece at the end. We'll use the letter "L" to label these points.

3. Some points are SOMETIMES WIN points.

This means that if you start by removing a piece from one of these points, you may win or you may lose, depending upon how you play. We'll use the letter "S" to label these points.

For the three-point version, you already know how to label the sites:

W   L   W

The four-point version would look like this:

L   W   W   L

Strategies and Mastery

At this point, you should be ready to try some of the more complicated versions of the game:

1. First, decide what version you want to play. You may want to start with a small array, 2 x 3 or 3 x 3, for instance.

2. Play this version several times.

3. See if you can label which starting points are L, W, or S for this version. See the box at left for tips on how to figure this out and how to record your findings.

4. Once you've mastered a simple array, you can move on to a larger one. Then see if you can figure out the Ls, Ws, and Ss for that one.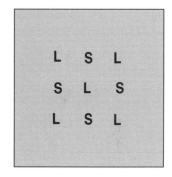

The more you play, the more you'll learn about the strategy of the game, such as where to begin (or not begin), how to move, and how to think ahead about what pieces to move next. What's also interesting is that you can learn to predict whether you will win or lose a game.

You will have mastered any version of this game when you can correctly label the points on the game board with a W, L, or S. Even if you can't always win when you start with an S point, you'll have mastered the game if you know for sure that your starting point is an S.

Here are four one-dimensional arrays, with their W, L, and S points labeled:

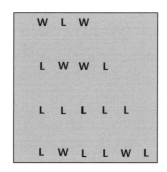

```
W   L   W

L   W   W   L

L   L   L   L   L

L   W   L   L   W   L
```

As you discovered when you played the 1 x 3 and 1 x 4 arrays, mastery of the smaller versions of the game can be quite simple. It provides an opportunity for you to go on to more complex games and look for generalizations.

The principles you discover may give you theories about other versions that seem similar. For example, if you're convinced you've mastered a 4 x 4 game, you might decide that the same principles will apply to any game that has a square playing area, or any game that has an even number of points along one side.

Take the time to test your theories. You'll find that sometimes they're right—but sometimes they aren't.

You may come to the conclusion, for example, that mastery of a 4 x 4 game yields:

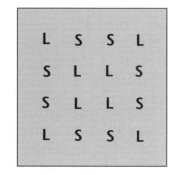

```
L   S   S   L
S   L   L   S
S   L   L   S
L   S   S   L
```

So you might also be tempted to conclude that a 3 x 3 game should yield an abbreviated version, like this:

```
L   S   L
S   L   S
L   S   L
```

Or for a 5 x 5 array, an expanded version, like this:

```
L   S   S   S   L
S   L   S   L   S
S   S   L   S   S
S   L   S   L   S
L   S   S   S   L
```

But both of those generalizations are wrong. You can have an interesting time trying to figure out why, and what the correct W, L, and S points are.

So how do you decide whether a starting point is really a W, L, or S? What if you play a game lots of times, each time starting from the same point, and you don't win? Is that point an L, a sure loser, or an S that you haven't yet been able to win from?

With complicated versions of the game, you probably won't be able to come up with accurate conclusions based just on what you can demonstrate on a case-by-case basis. In general, to be able to have confidence in your conclusions, you would need to investigate what constitutes a mathematical proof.

Mastering Madagascar Solitaire

Once you've mastered some smaller versions, try playing the Madagascar version of the game, shown on page 27. It has 37 points, so it's a lot more complicated—and fun to experiment with. But it's not easy to win. In fact, there are some starting positions from which it is impossible for you to win.

Try setting up the game board and playing a few times. When you first start playing, don't be discouraged if you have six or seven pieces left at the end of the game. It takes a lot of practice to learn how to win. But here's one hint that will improve your play: You can't win unless you make moves that keep the remaining pieces as close together as possible.

Since the game is hard to win, it won't be easy for you to know which category (W, L, or S) any particular beginning point is in. If you play for a while, you may discover that there are some points you think you can label as L—if you start from that point, no matter how well you play, you can't win.

Here's the 37-point Madagascar Solitaire board labeled with the three categories of beginning points. What do you notice about this diagram?

Did you notice that there are no points labeled with a W? In a game this complicated, there are no points that will guarantee you will win. There are only starting points from which you MUST LOSE, and there are starting points from which it is possible for you to SOMETIMES WIN, if you play well.

Playing with the Game

You may also want to experiment with the rules of the game. For example, instead of winning when you leave only one peg, you might decide to play a version of the game where leaving a particular pattern is the goal—like the **T** shape shown here, which was suggested at a workshop for teachers in Tanzania.

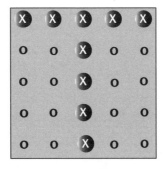

The basic game is so simple that your imagination is the only limit on the ways you can experiment and explore.

This activity was developed by Robert Lange.

Making Connections

- Have you ever played any other games that involve moving or jumping over the playing pieces? Which ones? How are they like or unlike Madagascar Solitaire?

- Can you use the strategies you've learned playing Madagascar Solitaire in any other games you play?

- Ask your parents or grandparents what kinds of games they played when they were children. Can you find out where those games originally came from? Are any of them games you still play?

Recommended Resources

Bell, Robert Charles. *Board and Table Games from Many Civilizations*. New York: Dover, 1980.

Cassidy, John. *The Book of Classic Board Games*. Palo Alto, CA: Klutz Press, 1991.

Tyson, Peter, and Russell A. Mittermeier. *The Eighth Continent: Life, Death, and Discovery in the Lost World of Madagascar*. London: Avon Books, 2000.

Van Delft, Peter, and Jack Botermans. *Creative Puzzles of the World*. Emeryville, CA: Key Curriculum Press, 1995.

COUNTING AND CALENDARS

4

QUIPUS

The Inca Counting System

QUIPUS The Inca Counting System

What if you had no way to write numbers? What if you had a collection you wanted to trade, but you had no way to count your dolls or old bottles or baseball cards? How could you let someone else know how many items you had, and what you thought they were worth? The people of the Inca empire created a unique solution to this sort of problem: a record-keeping device called a *quipu*.

The Empire of the Inca

The Inca empire, which thrived for centuries in the area that is now Peru, existed until 1535, when it was destroyed by the Spanish con-quistadores. It covered an area that stretches almost 2,500 miles—from what is now southern Ecuador to central Chile. It was made up of different groups of people, but everyone spoke *Quechua* (pronounced "ketch-wah"), a language still spoken by several million people in Peru, Bolivia, and parts of Chile today.

The Inca had no written language. What we know of them comes from the written records of Europeans who conquered and destroyed the Inca empire in the sixteenth century.

According to these documents, the Inca had a highway system that connected all parts of the empire and allowed them to keep the records for the entire kingdom up to date. The Inca roads were so well built that many of them still exist today.

WHAT'S IT ALL ABOUT?

In this chapter, you'll learn about the Inca and how they counted and kept records. Then you'll make a quipu of your own. You'll also discover:

- How the Inca kept records of numbers without writing anything down
- How you can "write" a number—any number—using only string
- How archaeologists figured out how to read quipus
- How the Inca counting system was like—and unlike—our own

Is there anything special I should know?

- This activity is recommended for ages 10 and up
- You'll need to understand place value in the base-ten number system
- To make your own quipu, you'll have to learn to make the knots shown on page 43

- This activity works best if it's done in small groups, since different people will see different things about the way the quipu knots are organized

How much time will I need?

- Allow about two hours to learn about quipus, and another hour or two to make one of your own

What materials will I need?

- The diagram on page 41
- Two different weights of light rope or heavy string, such as clothesline rope and packing string, or packing string and thick yarn (if you wish, you can use different colored string in your quipu, too)

To govern well, the Inca leaders needed to get information from all parts of the empire. The roads made this possible. Along the roads were houses where the best runners of the empire waited for their turn to carry news from one place to another. As a runner came near a house, he would call out to the next runner, who would pass the message in the same way a few miles later, at the next runner's house. In this way, information could be carried for hours or days by runners who were always fresh.

The Inca leaders received messages and sent instructions this way every day. Messages might include details about what was in a city's storehouses, what taxes were owed or collected, the number of people in each part of the kingdom, or how much metal was being mined in a particular region. This much information could not be memorized and passed on from runner to runner without making mistakes. Thus messages had to be put in a form that was clear to understand and easy to carry.

How did they do it? Since the Inca had no written language, they kept track of this information—and much more—using knotted strings called *quipus*.

Unraveling the Knots

Quipus (pronounced "kee-poos") are lengths of rope or cord that have different kinds of knots tied into them. The knots are arranged in various patterns, and sometimes the cords are in different colors. Each color, type of knot, and pattern of knots has a meaning.

Quipus were made by specialists—people of high rank who were specially trained in the art of making and reading quipus. Some of these people served the emperor directly. Some traveled, visiting settlements and sending back reports. Still others were brought to the capital, at Cuzco, from all over the empire to learn the art of the quipu. These select quipu-makers then returned home to serve as communications experts representing their districts. People studied as long as two years to learn the skills of the quipu and were highly respected.

About five hundred quipus have been found in Inca graves. By studying them closely, anthropologists have concluded that the knots on a quipu represent numbers. So even though there is no record of Inca writing, there is a way to figure out how the Inca counted. In fact, the information a quipu contains is organized in a way similar to what we call a database today.

Think Like an Archaeologist

Like the archaeologists who first studied the quipu, the best way to discover its secrets is to look closely at one. Begin by examining the real quipu shown on page 37. What do you notice about the way it's constructed? ☉

Now take a look at the quipu in the photograph above. It's a sample

These illustrations and quotes come from a 1,200-page manuscript sent to the king of Spain in 1615. The author of the manuscript, Guamán Poma, was a noble Inca who joined the Spanish administration after the Inca were conquered. The manuscript is titled *Nueva crónica y buen gobierno,* or *New Chronicle and Good Government.*

THIS MAN HOLDS A STRETCHED-OUT QUIPU IN HIS LEFT HAND AND A ROLLED-UP QUIPU IN HIS RIGHT HAND. THE TEXT SAYS HE IS ONE OF THE "SONS OF THE HIGH NOBILITY OF THESE KINGDOMS WHO GAVE THEM THESE POSITIONS SO THAT THEY WOULD LEARN A TRADE, LEARN TO RECKON AND GIVE ORDERS . . . TO GOVERN THE LAND. . . . THOSE CALLED HONEST SECRETARIES CARRIED QUIPUS DYED IN VARIOUS COLORS."

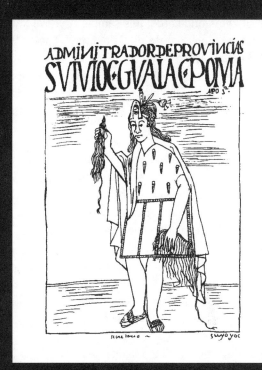

THIS IS A MAN WITH A QUIPU IN HIS LEFT HAND. THE AUTHOR SAYS HE IS ONE OF THE "PHILOSOPHER/ASTROLOGER INDIANS WHO KNOW HOURS AND SUNDAYS, AND DAYS AND MONTHS TO SOW AND REAP THE FOOD CROPS OF EACH YEAR . . . WHO KNOW ABOUT STARS AND THE RETURNING OF THE PATH OF THE SUN."

THIS MAIL RUNNER CARRIES A SMALL SIGN THAT READS "CARTA" (LETTER). HE IS CARRYING A QUIPU IN HIS RIGHT HAND. THE MANUSCRIPT SAYS THAT "THESE YOUNG MEN SERVED AS MESSENGERS . . . FROM ONE VILLAGE TO ANOTHER."

we made to show you the parts of a real quipu.

Although it may look like a tangle of knots and cords, this sample quipu actually contains familiar information. Once you learn how to read the knots, it will be easy for you to figure out the message.

What's in a Quipu?

Look at the diagram of our sample quipu on the next page. This diagram will help you see all the parts of a quipu. You can also use it to figure out the hidden message in the sample.

Look carefully at the diagram. What do you notice?

First, you'll notice a *main rope* that is thicker than the others. A number of cords—called *principal cords*—are tied onto the main rope. Thinner cords are attached to the principal cords. They are called *subsidiary cords*.

You may notice that part of the main rope curves down at the far left, and has a subsidiary cord attached to it. This part of the main rope and its subsidiary seem to have more knots than the other cords.

Now you can begin to figure out what this quipu means. Since you know that the Inca used quipus for counting, you might want to begin by counting, too. Start with the cords themselves. How many principal cords are hanging from the main rope?

Remember that the cord at the far left is part of the main rope, so there are 12 principal cords hanging from the main rope. Everything you discover about the quipu is a clue. What might this one mean?

Counting by the Knots

Before you can figure out what's being counted, you'll need to dis-

cover more information and find more clues. As you can see, all of the cords hanging from the main rope have knots on them. Where are the knots? How are they grouped? Find the symbols for the different kinds of knots on the right-hand side of the diagram. What do you notice about the different kinds of knots? Look for patterns. Look for places where patterns are broken, too.

As you can see, three different kinds of knots are on the quipu. Some are in groups near the bottoms of the cords; others are in the middle. On the subsidiary cord attached to the main rope, there is a group of knots near the top. In a few places, there are no knots where you might expect to see some.

Take a look at the locations of the different kinds of knots. On the top two levels of the cords, you'll find only single knots. They look like this:

On the bottom level, there are *figure-eight knots*, which look like this:

and *long knots* of many different sizes, which look like this:

For the long knots, each horizontal line tells how many times the rope is wound around itself. Look at the drawings on page 43 to see how the knots were made.

Why do you think the long knots would be made in different sizes?

There are a few clues in the quipu. Look at the bottom level of the diagram and find the shortest of the long knots there.

How many of the shortest long knots did you find? Where are each of those knots? (Count the principal cords from left to right.)

You've probably found one two-turn long knot on the second principal cord, and another two-turn long knot on the twelfth principal cord. Does this give you any clues?

Now let's look for the longest long knot on the quipu. What do you notice about that knot?

You'll find that the longest long knot has nine turns. And, as you can see, the long knot with nine turns is on the ninth cord. Is there a pattern here? Study the principal cords and see what you discover.

How many turns are on the long knot on the eighth principal cord? On the seventh principal cord? Track your way back to the smallest long knot, with two turns, on the second principal cord. You've discovered how to count from 2 to 9 on a quipu!

But what about the number 1? Where would you expect it to appear on this quipu?

Finding the number 1 should be simple enough. Look at the first principal cord. If the pattern you've been following holds true, then the figure-eight knot on the first principal cord must represent the number 1, since it comes before what you read as the number 2.

On Beyond Ten

Now you know quite a lot: A figure-eight knot is a 1, and the long knots count 2 to 9, depending on the number of turns they have. But what happens after that? There are no long knots with ten, eleven, or twelve turns. Why do you suppose the long knots don't just get

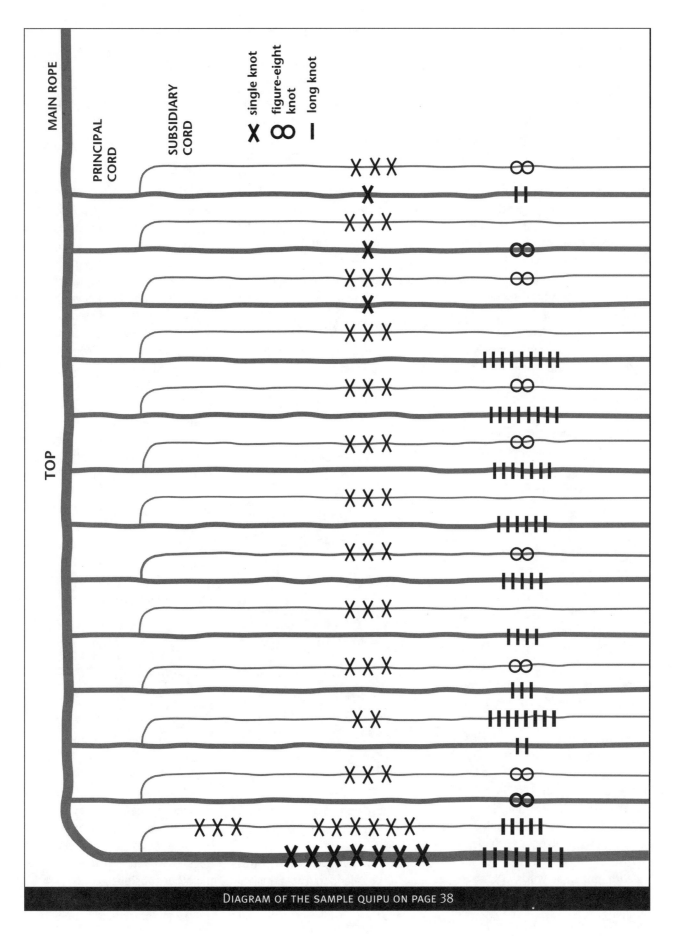

longer and longer to show higher and higher numbers? 🔵

Think about what happens after the number 9 in your own number-writing system. Picture the numbers 1 through 9 written in a column, one below the other. What changes when you write the numbers after the 9? 🔵

People in different cultures have different ways of writing numbers. In the United States, we group what we are counting by ones, tens, hundreds, thousands, and so on. To record data, we write down our symbols for numbers in different positions along a line that runs from left to right. When we look at any number, we know at once what the symbols and their positions mean.

For example, 1,776 means:

one thousand: 1 in the far left position, the thousands place

seven hundreds: 7 in the next position, the hundreds place

seven tens: 7 in the next position, the tens place

six ones: 6 in the far right position, the ones place.

The Inca would represent the number 1,776 in a different way. Follow the pattern you found from 1 on the first principal cord up to 9 on the ninth principal cord. How do you think 10 is shown on the quipu? How is this way of showing 10 like our way? 🔵

Notice that there is no knot on the bottom level of the tenth principal cord—but there is a single knot on the middle level. The single knot means that there's one 10; the empty space at the bottom is a zero—there are no ones on this cord.

Now look at the eleventh and twelfth principal cords. Do those knots fit the pattern you've been building? 🔵

The eleventh cord has one single knot at the middle level, 10, and one figure-eight knot at the bottom level, 1. It shows the number 11.

The twelfth cord has one single knot, 10, and a long knot with two turns, 2, to represent the number 12.

You've discovered that the Inca counted using 10 as a base, the way we do. They tied the knots on their quipus at different positions from bottom to top along the cords, just as we write our digits in different positions from the smallest on the right to the largest on the left.

Solving the Puzzle

Now that you know what all the individual knots and their levels mean, you can read the numbers on all the principal and subsidiary cords of this quipu. 🔵

As you count, you'll get the series of numbers below:

Principal Cords	Subsidiary Cords
1	31
2	28
3	31
4	30
5	31
6	30
7	31
8	31
9	30
10	31
11	30
12	31

Do these numbers look familiar? What do we count in 30s and 31s, with one 28 thrown in?

You may have guessed that this quipu is a calendar. The principal cords count the 12 months of the year, and the subsidiary cords count

the days of each month. (Remember that this is a sample made to demonstrate how quipus were used. The numbers on this quipu come from the calendar we follow today, not from the Inca.)

One Last Mystery

Now that you know what this quipu means, you have just one mystery left: the part of the main rope that hangs down on the left. Let's go back and figure out what it's for.

Look at the quipu diagram again. You already know that the bottom level of knots stands for the ones and the middle level of knots stands for the tens. What do you think a third level of knots might represent? What if there were fourth and fifth levels? What would those knots represent? 🔵

You probably figured that the third level of knots counts the hundreds, and that fourth and fifth levels would count thousands and ten thousands, and so on. With this in mind, try reading the numbers on the left-hand part of the main rope and its subsidiary cord. 🔵

You'll find that the knots represent these numbers:

Main Rope	Its Subsidiary Cord
78	365

(If you didn't get these numbers, go back and try counting the knots again.)

What do you think these numbers could mean? 🔵

The number 365 is familiar—it's the number of days in a year. How does 365 relate to the numbers on the other subsidiary cords? 🔵

If you add the number of days in the 12 months, you get 365, the number of days in a year. So the first subsidiary cord gives the sum of the numbers on all the other subsidiary cords.

Now look at the part of the main rope with knots that represent the number 78. The number 78 is not familiar the way 365 is. But could it also be a sum?

If you add the numbers on all the principal cords (1 + 2 + . . . + 12), you get 78. That's how the Inca made their quipus: One special cord on the side summed up the rest of the cords. It was like double-checking the numbers in the quipu.

Do-It-Yourself Quipus

Now that you know how to count like an Inca, you can make your own quipu. The first step is to learn to make the three different kinds of knots shown on this page.

Try making each of the different knots a few times, and see if you can read the numbers easily. Then choose a three-digit number and show it on one string. Try placing the ones, tens, and hundreds three or four inches apart along the string. (If you need to move a long knot, you can loosen it by pinching the half-loops at its ends and pulling them apart along the string with your fingernails. Then you can slide the whole knot in the direction you want to move it.)

If you're working in a group, you may want to see if someone else can read your three-digit number.

Making Your Own Special Quipu

Think of some numbers you would like to record. They could be the ages of your friends or family, the amount of time you spend in school or at work each week, the number

HOW TO MAKE QUIPU KNOTS

Single Knot

This is a simple knot. Be careful not to tighten it too much. You want it to keep its shape and be easy to identify.

Figure-Eight Knot

Follow the drawing to tie the figure-eight knot. As you gently tighten it, notice how different it looks from the single knot.

Long Knot

The long knot can contain from 3 to 10 turns of the string around itself. As you pull on both ends to tighten the knot, it will reorganize itself in a surprising way. Once the knot is tight, you'll find that it shows one less turn than the number of turns you started with. To make a 2-turn long knot, you'll need to wind it with 3 turns; to make a 9-turn long knot, you'll need to start with 10 turns.

of items in your collection, the important dates in your year—anything you choose.

Now record those numbers on a quipu of your own, attaching principal and subsidiary cords to a main rope in any way you wish. When you're finished, you can pass your quipu to a friend, but don't tell that person what you have recorded. Can your friend read the numbers on your quipu? Can he or she figure out what those numbers mean?

What could you do to make your quipu more readable?

Once you know how to make and read a quipu, you'll be able to appreciate various ways of recording data.

This activity was developed by Maurice Bazin and Modesto Tamez.

Making Connections

- What kinds of things do you need to keep track of by counting? What does your family need to keep track of? What methods do governments use to count and keep records? Do you think everyone needs a counting system?

- Suppose you had no writing system. How would you keep track of important information? What materials would you use?

Recommended Resources

Ascher, Marcia, and Robert Ascher. *Mathematics of the Incas: Code of the Quipu.* New York: Dover Publications, 1997.

Bernard, Carmen, and Paul G. Bahn. *The Incas: People of the Sun.* New York: Harry N. Abrams, 1994.

Guamán Poma de Ayala, Felipe. *Letter to a King.* New York: E. P. Dutton, 1978.

5

COUNTING LIKE AN EGYPTIAN

Math in Ancient Egypt

COUNTING LIKE AN EGYPTIAN Math in Ancient Egypt

With its mysterious sphinx, pyramids, gods, and mummies, ancient Egypt stirs our imaginations. Besides a rich artistic and spiritual history, ancient Egyptians created hieroglyphics, a complex system of writing and calculation that we still study today.

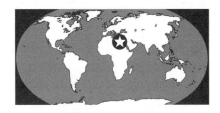

A Little About Egypt

Northeast Africa's Nile River region has been populated for more than six thousand years. Dynasties of kings ruled the civilization we know as ancient Egypt continuously from 3000 B.C. until Alexander the Great's conquest of the area in 332 B.C.

The three huge pyramids at Giza were built around 2550 B.C., pre-sumably as burial sites for individual kings. It was a rich period of creativity: The architects and scribes of the time were revered, and mathematics prospered.

Only two ancient Egyptian mathematical manuscripts—written on long, thin rolls of material made from the papyrus plant—have been recovered. They contain specific mathematical problems and their solutions, such as the volume of a truncated pyramid and the volume of a cylinder. But monuments and burial sites also bear mathematical data related to astronomy and the Egyptian calendar.

Most of what we know about ancient Egypt has come from solving the mystery of hieroglyphics, a system of writing that looks like a series of pictures. Until nineteenth-century scholars were finally able to "break the code" of the hieroglyphics, they had to rely on information from Greek texts, which only discussed Egyptian life at the very end of the great dynasties.

Like English, the Egyptian language was mostly made up of syllables. Each syllable was represented by a hieroglyphic symbol—a bird, a knot on a rope, an eye, and so on. As a result, the drawings do not

WHAT'S IT ALL ABOUT?

In this activity, you'll follow the same steps an archaeologist might use to figure out some Egyptian hieroglyphics. Along the way, you'll explore:

- How you can use simple observation to begin translating a text in a mysterious language
- How the counting system of the ancient Egyptians was like ours, and how it was different
- Why a counting system might need fractions

Is there anything special I should know?

- This activity is recommended for ages 12 and up
- You can do this alone, but it's easier—and more fun—to work with others in a small group

How much time will I need?

- Allow 2 to 3 hours

What materials will I need?

- Copies of the Egyptian text translated by Théophile Obenga, on page 50

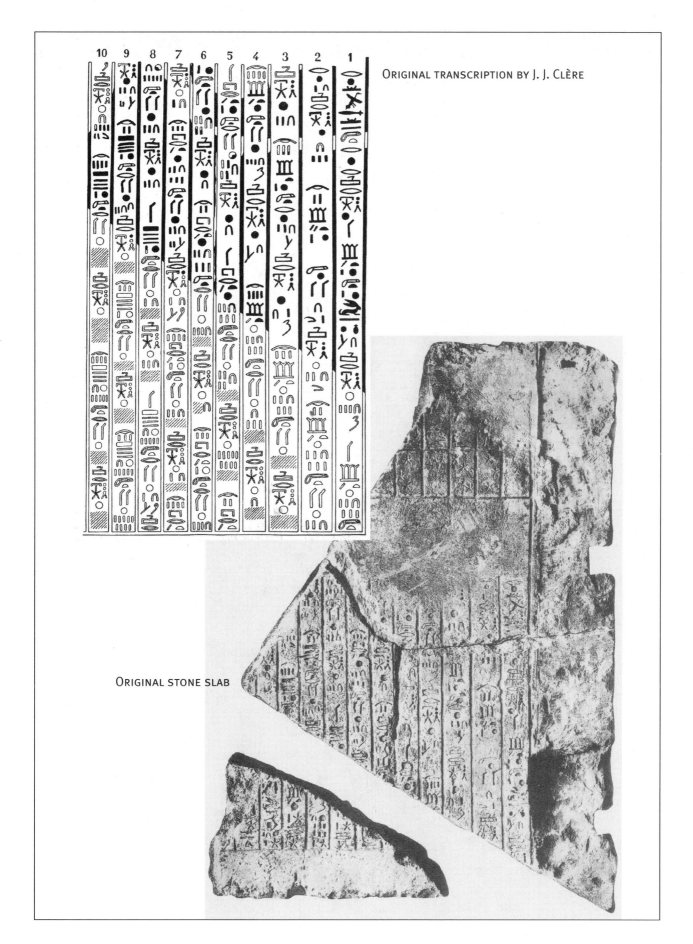

ORIGINAL TRANSCRIPTION BY J. J. CLÈRE

ORIGINAL STONE SLAB

Modern transcription by Théophile Obenga

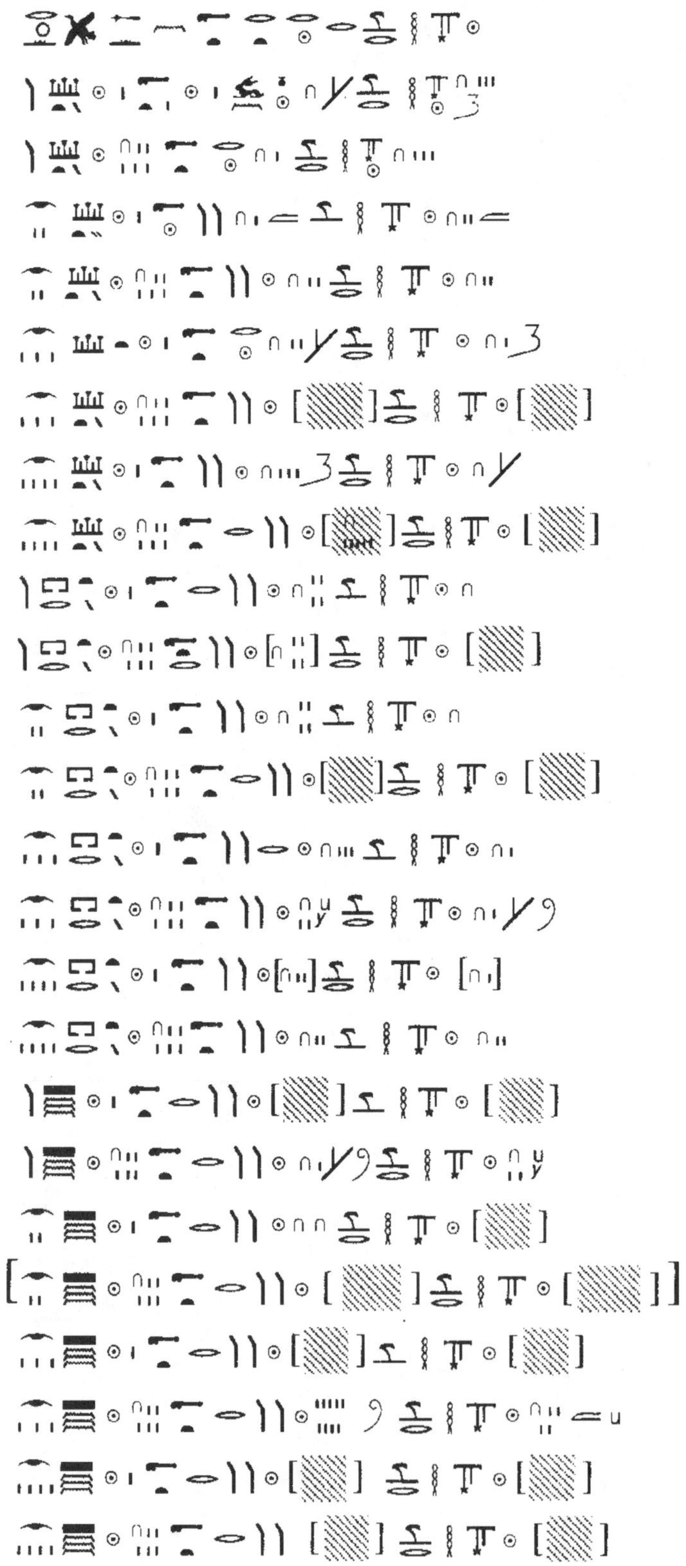

convey meaning by themselves. Like the letters of the English language, each symbol represents a single sound. The hieroglyph ⏥, for example, which appears repeatedly in the text you'll be working with, represents the consonant sound *R*.

About the Egyptian Text

On page 49 you see a picture of a stone slab that bears the inscriptions you'll be working with in this activity.

The stone was found covering the top of an abandoned well in the ruins of the ancient Egyptian city of Tanis. It was probably part of a wall: You can see the structural slots cut into one side. If you look at the stone carefully, you will also see that some of the inscriptions are readable, while others are not.

One way to begin to unlock the secrets of a stone like this is to copy the carved symbols onto paper, so you can study them more easily. The first archaeologist to work with this stone, the French Egyptologist J. J. Clère, transcribed it as shown on page 49. The existing, readable parts of the slab appear outlined in black. The missing or unreadable portions are shown as gray outlines. Clère made educated guesses about what these portions might contain. He did not reconstruct the crosshatched areas.

The Egyptians always read from the top to the bottom of a column. Sometimes, within each column, they wrote from left to right, but other times, they wrote from right to left. Many documents mixed the two ways of writing. How could you know which way to start?

Look closely at Clère's original transcription. Can you find two recognizable symbols in the first column of the text? 🌀

You probably found a flying bird near the top of the column and a rabbit about halfway down. Notice that both face to the right. The carved animals actually told the Egyptian reader which way to read. An animal facing to the right, like the ones here, told a person to read the column from right to left; an animal facing left meant that the column should be read from left to right.

Finding the Patterns

For this activity, you'll be using a more modern translation of the same stone slab, done by a Congolese philosopher named Théophile Obenga. His transcription, as you can see on page 50, is very different from Clère's. Notice that he's turned the bird and the rabbit around so that they face to the left, instead of to the right.

How can you use Obenga's transcription to unlock the meanings of the symbols on the stone? Imagine again that you are an archaeologist. Since you are working with an unfamiliar document, you might want to start by looking for patterns.

Obenga rearranged the text so that it reads from left to right, the way you would read an English text. Each line is like a sentence, making it easier to see patterns since the writing is arranged the way we're used to reading.

Try looking at a copy of the text from a distance. Squint a little. Both of these methods may make it easier for you to see patterns instead of individual symbols.

Finding groups that go together is one way to begin. Do you see groups of sentences that look similar, or that seem to go together? Draw horizontal lines to divide the text into blocks of related sentences. Use a pencil so you can make changes later, if you want. 🌀

Your text should look something like the example on page 53. There are seven sections in the text. If you divided it up differently, go back and take a look at the patterns again. This time, focus your attention on the symbols at the beginning of each line.

If you start from the top and work down, you may notice that the very first line is different from the other lines. For one thing, it shows one of the text's only recognizable images—the flying bird (the rabbit shifted to the next line). It's shorter and simpler than the other lines, and it only has two bull's-eyes in it—the symbols that look like this: ☉ . Most of the other sentences have three of these bull's-eye symbols. Why do you think this first line might be so different? 🌀

You might have thought of the idea that the first line is a title for the rest of the text. You can't know that for sure, but it seems like a good possibility. So for now, ignore the first line and concentrate on the rest of the text.

Even though you've grouped similar sentences, the text probably still looks like a hodgepodge of strange marks and symbols. Try dividing the page into even smaller chunks. Notice that the bull's-eyes, ☉ , which seem like natural separators, line up roughly from top to bottom. The first one in each line is near the beginning, the second one in most lines is near the middle, and the third one is near the end.

Use your pencil to connect the bull's-eyes vertically, like a dot-to-dot picture, and divide the text up into four jagged vertical columns.

When you're done, your text should look like the example on page 54. 🌀

What's in a Word?

Now that you've broken the text down into manageable chunks and identified some of the individual symbols, what else can you discover? Some of the symbols repeat at the beginnings of lines, and the bull's-eyes repeat in all the lines. What other repeated symbols can you find? ●

Other symbols that are used over and over again look like open eyes ◡ or closed eyes ◠. There are also symbols that look like chains ⅄, pipes ━, and hockey sticks ⅃⅃, and some that look like upside-down magic wands ⊤.

See if these repeated symbols give you more information about the text. Look for groups of symbols and notice how they're arranged. Do some symbols show up next to each other a lot? What patterns can you see? ●

You've probably noticed that some symbols often appear together in the text. For example, the chains ⅄ in the third column always have the magic wands ⊤ just to the right of them. In the second column, pairs of hockey sticks ⅃⅃ always go with pipes ━ and half circles ◣, and often with closed eyes ◠.

Could these groups of symbols be words? We know that the symbols in the Egyptian text stand for sounds, as our English letters do. Perhaps groups of symbols that always go together are like the letters or syllables that make up a word. This makes sense when you see that the same "words"—the same groups of symbols—keep appearing throughout the text.

Looking for the Numbers

Scholars discovered that Egyptian hieroglyphics represented numbers as well as words. Could there be both words and numbers on the

stone slab? How could you tell the difference?

One way to think about this is to imagine what would happen if an alien landed on a baseball field on earth in the middle of a game. The alien might notice a huge board with strange characters on it. Is there some organization to the display that might give the alien clues to figure out what the marks represent?

Inning	1	2	3	4	5	Total
North City Giants	0	II	IIII	0	0	IIIII I
South Town Pirates	I	0	0	II	I	IIII

The alien is likely to notice a difference between the strange, complex markings (the names of the teams) and the simpler-looking tally marks (the scores).

The alien might not be able to read the words or figure out why the humans were whacking at a round ball with a wooden stick, but our extraterrestrial visitor might at least be able to figure out that the simple repeated lines and groups of lines meant the humans were counting something.

As you look at the Egyptian text, you can use this same technique to tell the difference between hieroglyphic numbers and words. Can you find some simple symbols that look like tally marks? ●

At the beginnings of many of the "sentences" in the Egyptian text, there are a number of small, vertical lines ı ı ı. If you look carefully, you'll find that there are also groups of tally marks on the left side of some other columns, just to the right of the bull's-eyes. It looks like something is being counted here. What could each vertical bar represent? ●

Because the tally marks are simple, and come in different-sized groupings (like the tally marks on the scoreboard), it seems likely that each mark represents the number 1.

Weird 1s

You can test this assumption by looking at the tally marks in each line on the manuscript. What do you see when you look down the first column of the text? ●

The first line is probably a title, so don't worry about that for now.

The second and third lines each start with a single hockey stick ⅃. But the fourth and fifth lines each start with two tally marks—the number 2. The next two lines start with the number 3, and the two lines after that start with the number 4. This eight-line pattern is repeated again in the last eight lines.

Look at the hockey sticks at the beginnings of lines 2 and 3, 10 and 11, and 18 and 19. Could they also represent the number 1? They look as if they should. Then why are there hockey sticks in the second

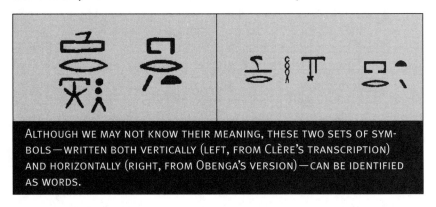

ALTHOUGH WE MAY NOT KNOW THEIR MEANING, THESE TWO SETS OF SYMBOLS—WRITTEN BOTH VERTICALLY (LEFT, FROM CLÈRE'S TRANSCRIPTION) AND HORIZONTALLY (RIGHT, FROM OBENGA'S VERSION)—CAN BE IDENTIFIED AS WORDS.

column, right in the middle of some more complex symbols that might be words? Maybe the Egyptians had more than one way to show numbers—like the tally mark (l) and Arabic numeral (1) used on the scoreboard on page 52.

Beyond the 1s

If a single tally mark corresponds to the number 1, you can use them to count the simplest numbers. But how far can you count using these marks? What is the biggest grouping of tally marks you can find in one line of the text? ◖

You should find that the biggest group has nine tally marks: In our example on page 54, it's in column three, in the third row up from the bottom.

If nine tally marks represents the number 9, how did the Egyptians write the number 10? Maybe the symbol ∩ which is next to many of the individual tally marks, represents the number 10. This would mean that the Egyptians used a base-ten counting system, like ours.

Take some time to see where the number-10 symbol appears. Can you see any patterns—or breaks in a pattern? ◖

You probably noticed that the number-10 symbol appears sometimes with the 1s, and sometimes by itself. Sometimes it's written next to the 1s, and sometimes it's above them. From that information, you can figure out that although the Egyptians also counted in base ten, their system did not use position to give value, the way ours does.

Putting It All Together

Now you should be able to read all the numbers in the text made with the number-10 and number-1

DRAWING HORIZONTAL LINES IS ONE WAY TO GROUP SENTENCES TOGETHER AND DIVIDE THE TEXT INTO SMALLER BLOCKS.

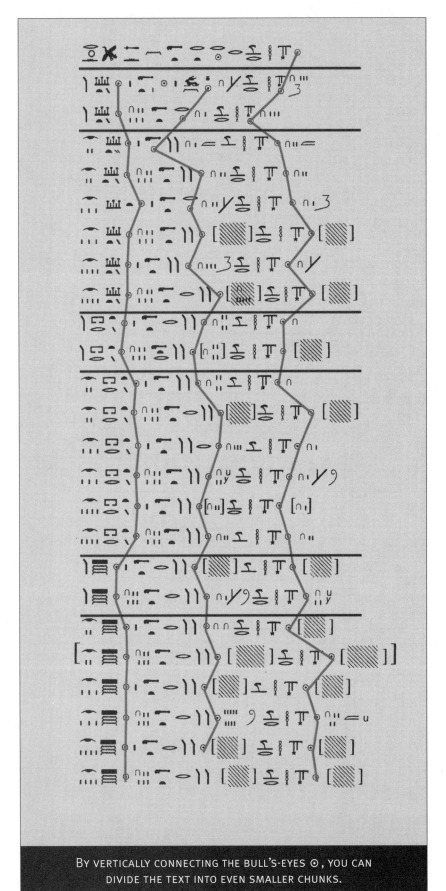

symbols. This is a good time to create your own transcription of the text. Get a piece of notebook paper and write down all the numbers you can translate. Use one line of the notebook paper for each line of the Egyptian text, and divide your whole table into four columns. You should end up with a table that looks like the one on page 55. Your table will make it a lot easier to see how the numbers are arranged.

Now that you have something more familiar to work with, you can start to think about what these numbers might mean.

Take a look at the first column. If you assume a hockey stick is also a 1, this column begins with a simple pattern of 1, 1, 2, 2, 3, 3, 4, 4. Then the same pattern is repeated. The second column is also regular: It jumps back and forth between 1 and 15.

The first and second columns seem to go together. They seem to count the first and the fifteenth of one thing, the first and the fifteenth of a second thing, the first and the fifteenth of a third thing, and so on.

The table has twenty-four lines. What if you look at it as twelve sets of two lines each (one set has 1s, one set has 15s)? Does this remind you of anything? What common event is counted in units of twelve and might be divided into the first and the fifteenth?

You may have been reminded of a calendar, since there are twelve months in a year. People who have jobs and are paid twice a month might recognize the repeating pattern of the first and the fifteenth and think: Aha! Payday! This text marks the first and the fifteenth of one month, the first and fifteenth of the next month, and so on.

Counting and Calendars

If these 24 lines represent the first and the fifteenth days of twelve 30-day months, that fits with what we know from other texts. The Egyptian year had 360 days divided into twelve equal months, plus five extra days. So you can assume that this text is about something that happened twice a month during the year.

Now that you have a working theory about the first two columns, you might want to think about the information in the third and fourth columns. Using your table, look down each of these columns, then look at the two together, and see if you can discover any patterns there. ◐

You may see that (except for one oddball 9), the numbers in these two columns are always larger than 10, and most are close to the number 12. You might also notice that as the numbers in column three go up, the numbers in column four go down. But after twelve lines, this pattern changes: The numbers in column three begin to go down, and the numbers in column four begin to go up.

Cryptographers (people who break codes) sometimes add numbers together or subtract them from each other to see if they can discover hidden patterns. You might try that here—adding or subtracting the numbers in columns three and four on each horizontal line to see what you get. ◐

Subtracting the numbers horizontally doesn't give you much of a useful pattern. But if you add the numbers horizontally in columns three and four, you'll find that many of the lines add up to 24. What can you think of that is counted in units of 24, and divided into two groupings? ◐

TITLE LINE			
1?	1	10	13
1	15	11	13?
2	1	11	12
2	15	12	12
3	1	12	11
3	15	?	?
4	1	13	10
4	15	?	?
1?	1	14	10
1?	15	14	?
2	1	14	10
2	15	?	?
3	1	13	11
3	15	12	11
4	1	12	11
4	15	12	12
1?	1	?	?
1?	15	11	12
2	1	(20)*	?
2	15	?	?
3	1	?	?
3	15	9	14
4	1	?	?
4	15	?	?

*The 20 is so different from the other numbers that it's probably an error.

ONCE YOU KNOW HOW TO FIND THEM, FOUR DISTINCT COLUMNS OF NUMBERS BECOME EVIDENT.

TITLE LINE

1?	1	10✓	+	13 ⌐	= 23 + ✓ + ⌐ = 24
1?	15	11	+	13	= 24
2	1	11 ⌣	+	12 ⌣	= 23 + ⌣ + ⌣ = 24
2	15	12	+	12	= 24
3	1	12 ✓	+	11 ⌐	= 23 + ✓ + ⌐ = 24
3	15	?	+	?	= ?
4	1	13 ⌐	+	10 ✓	= 23 + ⌐ + ✓ = 24
4	15	?	+	?	= ?
1?	1	14	+	10	= 24
1?	15	14	+	?	= ?
2	1	14	+	10	= 24
2	15	?	+	?	= ?
3	1	13	+	11	= 24
3	15	12 u ✓	+	11 ✓ ?	= 23 + u + ✓ + ✓ + ? = 24
4	1	12	+	11	= 23*
4	15	12	+	12	= 24
1?	1	?	+	?	= ?
1?	15	11 ✓ ?	+	12 u ✓	= 23 + ✓ + ? + u + ✓ = 24
2	1	(20)	+	?	= ?
2	15	?	+	?	= ?
3	1	?	+	?	= ?
3	15	9 ?	+	14 ⌣ u	= 23 + ? + ⌣ + u = 24
4	1	?	+	?	= ?
4	15	?	=	?	= ?

NOTICE THAT, ON THE LINES WHERE COLUMNS THREE AND FOUR ADD UP TO 23, THERE ARE ADDITIONAL SYMBOLS NEXT TO EACH OF THE NUMBERS. ONCE YOU HAVE ALL THE INFORMATION IN ONE PLACE, YOU CAN BEGIN TO SEE THE RELATIONSHIPS BETWEEN THE NUMBERS AND THE SYMBOLS.

You probably recognized that the number 24 could stand for the number of hours in a day. If that's the case, then some of these hours are daytime, and some are nighttime.

In most parts of the world, day and night are both close to 12 hours long. Archaeologists know the Egyptian words for "day" and "night." They look like this:

DAY	
NIGHT	

Can you find them in the text?

These words appear regularly in columns two and three.

Finding the Missing Pieces

All of this makes sense for the lines that add up to 24. But some of the lines only add up to 23. Why are they different? There's no such thing as a 23-hour day.

Take a look at the Egyptian text again. Notice that on the lines which add up to 23, the numbers are accompanied by small symbols that look like ⌣ or ⌐ or ? or ✓ or u. Add these symbols to your handwritten table. Do you notice any new patterns?

Your table may now look like the one at left. As you can see, the extra symbols appear only on the lines that add up to 23. The lines that add up to 24 have nothing added to their numbers. What could the extra symbols represent?

You may have figured out by now that the extra symbols are fractions. When you add them together, along with the number 23 on each line, you get a total of 24.

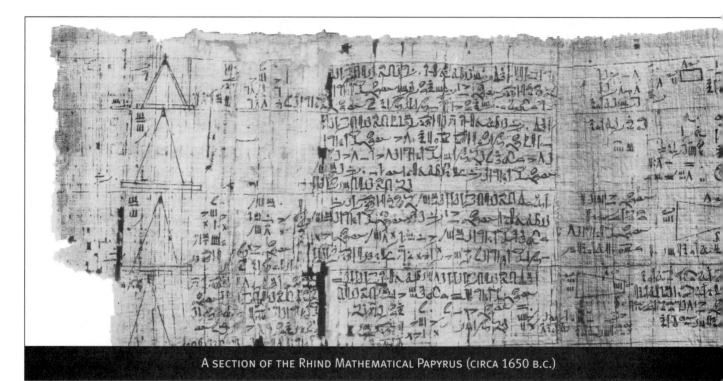

A SECTION OF THE RHIND MATHEMATICAL PAPYRUS (CIRCA 1650 B.C.)

DOUBLING DOUBLES: Egyptian Multiplication and Division

The ancient Egyptians' methods of multiplication and division were based on a process of doubling numbers over and over again. To multiply 7 by 13, for instance, they multiplied 7 by 2, then multiplied that number by 2. They did this over and over until the multiplier became larger than 13. In this example, they would stop at 7 x 8, since 16, which would be the next multiplier, is bigger than 13.

They transcribed the process like this, starting with 7 x 1:

\1   7
 2   14
\4   28
\8   56

When they stopped doubling, they put slash marks next to the multipliers that added up to 13. They got the result of the multiplication by adding up the numbers in the right-hand column of the lines with slashes. When the column is added up (7 + 28 + 56), it shows a result of 91: 7 x 13 = 91.

The Egyptians' method of dividing was closely related to their method of multiplying. To divide 91 by 13, they would ask, "By what must I multiply 13 to get 91?" To find the answer, they would multiply 13 by 2, again and again, like this:

1   13/
2   26/
4   52/

The series stops at the multiplier 4 because the next doubling would give 104, which is bigger than 91. In this case, they'd add all the numbers in the right-hand column to get 91, so there's a slash mark by each of them. When you add up the left-hand column, you get the answer: 91÷13 = 7.

Can you figure out what fractions are represented in the text?

Look at the third line on the chart on page 56. Notice that the two fractions on that line are the same. They look like this ⟨⟩. Since you know the fractions together must add up to 1, you can figure out that these two identical symbols must stand for the fraction $1/2$.

That means the third line represents day and night divided into $11\,1/2$ and $12\,1/2$ hours, respectively.

The other fractions in the text are $1/4$, $3/4$, $1/3$, and $1/6$. With the information here, however, you can't identify which symbol goes with which fraction. But notice that all the fractions in this text can be expressed as multiples of $1/12$. This suggests that the Egyptians divided each hour into twelve parts—that is, into five-minute intervals.

The Egyptians always used "unit fractions"—that is, fractions with a 1 in the numerator—to do their calculations. Archaeologists have found papyrus scrolls that give long lists of fractions expressed as sums of unit fractions. For example, on the back of the Rhind Mathematical Papyrus (an ancient Egyptian scroll inscribed with mathematical problems) is a table of the unit fractions that add up to $2/3$, $2/5$, $2/7$, . . . up to $2/101$.

For example:
$2/7 = 1/4 + 1/28$
$2/13 = 1/8 + 1/52 + 1/104$

You can find out more about ancient Egyptian math on page 57. The box "Doubling Doubles" shows the ingenious way the Egyptians solved multiplication and division problems.

Back to the Beginning

The processes you used to discover the meanings of the symbols in this text are a lot like the techniques used by the scholars who did the original translations. Today, scholars can read almost all Egyptian words. They know that the first line of this particular text means: "Knowledge of the difference between daytime and night."

As you have discovered, the text gives the lengths of day and night twice a month for a year. It also shows the names of the three Egyptian seasons, each of which were four months long. The words for the three Egyptian seasons look like this:

In the text, the months are labeled "first," "second," "third," and "fourth." The open-eye symbol ⟨⟩ is the word for "month."

The text also shows the equinoxes—the two times during the year when day and night are of equal length, each twelve hours long. The equinoxes are six months apart. In this 24-line text, the two equinoxes are twelve lines apart. Can you identify them?

Don't feel bad if you can't decipher everything on this stone: Neither can the archaeologists. But you may be surprised at how much you can figure out on your own just by looking closely and doing some basic math.

This activity was developed by Maurice Bazin and Modesto Tamez.

Making Connections

- The length of day and night changes throughout the year. How does this affect you? How does the change of seasons affect you? Do you ever have to wake up when it's dark or go to sleep when it's still light?

- Are you good at doing crossword puzzles and cryptograms? Have you ever played word games like Hangman or Scrabble? How are these games and puzzles like the work of scholars who study ancient languages?

Recommended Resources

Donoughue, Carol. *The Mystery of the Hieroglyphs: The Story of the Rosetta Stone and the Race to Decipher Egyptian Hieroglyphics*. London: Oxford University Press, 1999.

Gillings, Richard. *Mathematics in the Time of the Pharaohs*. New York: Dover, 1982.

Robins, Gay, and Charles Shute. *The Rhind Mathematical Papyrus: An Ancient Egyptian Text*. New York: Dover, 1990.

Woods, Geraldine. *Science in Ancient Egypt*. New York: Franklin Watts, 1998.

6

BREAKING THE MAYAN CODE

Mayan Math

BREAKING THE MAYAN CODE Mayan Math

If you found a book full of lines, dots, and mysterious-looking pictures, how would you begin to figure out what it meant? That was the problem facing archaeologists who discovered written records left by the Maya of Central America.

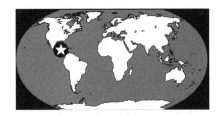

A Little About the Maya

The Maya prospered in an area ranging from southern Mexico and the Yucatán Peninsula through Belize, Guatemala, Honduras, and El Salvador. They first came to the area around 2600 B.C., and at the peak of their civilization were spread across an area of about 311,000 square kilometers (120,000 square miles). They lived in city-states, which were like tiny countries made up of a city and the land around it.

Skilled in the arts and sciences, the Maya flourished in the jungles of their homeland. They built roads to connect their cities. They were master architects, and their buildings are still considered amazing achievements. They were also astronomers who studied the cycles of the moon, earth, and other planets.

At the height of their culture, from the third century A.D. to the ninth century A.D., called the Classical Period, the Maya built large stone temples covered with stucco and colorfully decorated. Some of these impressive buildings remain today. If you climb the steep stairway to the top of the Temple of

WHAT'S IT ALL ABOUT?

In this activity, you will decipher a page from the Dresden Codex, one of the few Mayan books still in existence. By thinking like an archaeologist, you'll combine your mathematical abilities with some basic logic and trial-and-error investigation to figure out what the codex means. In the course of your explorations, you'll discover:

- How archaeologists have figured out what Mayan documents mean
- How the Mayan system of counting is like ours, and how it is different
- How Mayan beliefs were tied to their understanding of mathematics

Is there anything special I should know?

- This activity is recommended for ages 10 and up
- You can do this activity on your own, but it's much easier (and more fun) to work with others in a small group of 3 or 4

- In order to do this activity, you need to understand place value and number bases

How much time will I need?

- About 3 hours

What materials will I need?

- The portion of the Mayan codex shown on page 63
- Calculator (optional, but recommended)

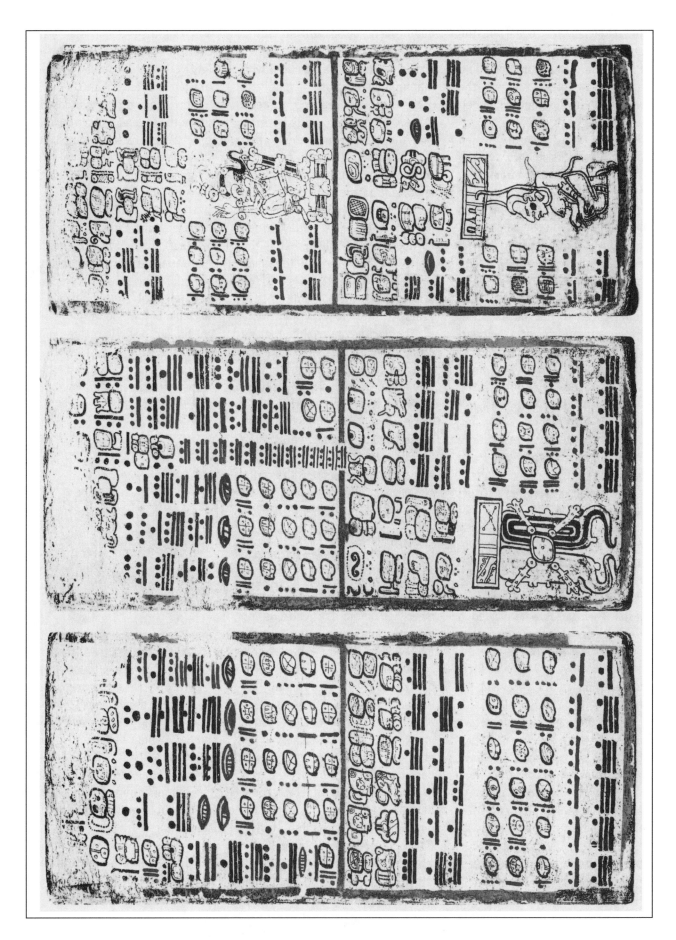

the Magician in Uxmal in the Yucatán, you'll get a bird's-eye view of the ancient city. From this point, you'll be struck by the way the Maya built in harmony with their surroundings.

In the third century, when Europe was in its infancy, the city of Palenque, on the Isthmus of Teuantepec, had a population of more than 100,000 people. This thoroughly modern city had its own drainage system and observatories, and buildings that towered 110 feet above the jungle floor.

In the 1500s, Spanish conquistadores invaded the Mayan cities. In their attempt to bring their version of civilization and religion to the Maya, the Spanish systematically destroyed Mayan books and documents that contained, according to the first bishop of Yucatán, "lies of the devil." As a result, little written information from the Maya survives.

The Mayan writings that still exist were carved on stone monuments or painted on pages of books that Western scholars call codices (singular, codex). Codices were made from pounded fig-tree bark treated with lime and covered with a thin layer of plaster. Their pages were painted in bright colors, folded accordion-style, and bound between pieces of wood. Of the few surviving Mayan codices, most are housed in European museums today.

From the rare documents we have, scholars have learned a lot about Mayan writing. The Maya had a number of different languages and a writing system of glyphs—symbolic pictures—that represented both words and syllables. Since Mayan glyphs can stand for both sounds and ideas, however, it's hard to know how to read each one. For example, a number could be written either with the number's symbol or with a picture of the god associated with that number—or both.

As scholars learned to read Mayan glyphs, they discovered that the Maya wrote about their lives and beliefs and kept extensive records of their possessions, important dates, and astronomical observations, many of which have proved to be valid today. When Europeans still believed that the world was only a few thousand years old, Mayan records alluded to life existing for millions of years.

The document you'll be looking at in this chapter is called the Dresden Codex because it found its way to the German city of Dresden. A portion of it is shown on page 63. The stains along one edge were caused by water damage during the firebombing of Dresden during World War II. Fortunately, the codex was saved.

Beginning with the Basics: Numerical Bases

Number bases have been invented by cultures throughout the world to meet their day-to-day needs. As you probably know, we count in base ten. Why do you think base ten is convenient? 🖐

The answer is simple: All humans—at some stage—count on their fingers. However, even though most people have the same number of fingers, not everyone counts in base ten. Every culture decides how to group things in order to count them. Some cultures count using only one hand, so their base is five. (In several African languages, the word for five means "hand full.") Other people count on both their hands and their feet, and use twenty as their base. Some Native Americans use base eight. Why eight? Count the number of spaces between your ten fingers.

Now imagine that you are visiting another planet where the intelligent beings are three-toed sloths. What bases do you think these beings would use for their sorting and counting? 🖐

The most likely bases for the sloths would be three (one foot), six (two feet), or twelve (all four feet).

If you understand the concept of using different bases, try practicing this important mathematical idea. The table below shows the number 459 broken down into three different bases. In each case, the 1s place is at the right, and the place values increase as you move to the left.

BASE 20:			
8,000s	400s	20s	1s
0	1	2	(Symbol for 19)

BASE 10:			
1,000s	100s	10s	1s
0	4	5	9

BASE 5:			
125s	25s	5s	1s
3	3	1	4

Why do you need a new symbol to represent the number 19 in base twenty? What other base-twenty numbers would you need to write with new symbols? 🖐

If you wrote "19" in base twenty, it would mean nine 1s and one 20 (which would be the number 29 in base ten).

In base twenty, every number less than 20 has to be written as a single digit in the 1s place. That's easy for the numbers 1 to 9. We already have symbols for those. But what about the numbers 10 to 19? They're two-digit numbers in base ten. So you'd have to create new symbols for all of them.

If you want more practice with different bases, choose two or three other numbers and write them in base twenty, base ten, and base five. 🌕

Which Way Is Up?

Take a look at the Mayan text on page 63. This text shows three pages of the Dresden Codex, one of the few existing examples of Mayan writing. Can you find some clues to help you figure out which way to read this writing? 🌕

Did you notice the position of the figure of the Mayan woman? This helps you tell which way is right side up. When archaeologists study a manuscript, they begin by using clues like that.

Now you can look for patterns in the Mayan writing. If you look at the pages from a distance, you'll see that the symbols on each page are in groups. If you still don't see a pattern, try squinting. 🌕

One thing to notice is that there are groups of symbols inside rectangles. In some rectangles, these symbols—called glyphs—are made up of elaborate designs with roundish borders. Other groups of symbols are made up of simple bars and dots. There are both numbers and words in this text. Which do you think is which? (Think about our own writing system: Do we have more symbols that make up words or more symbols that make up numbers?) 🌕

If you figured that the glyphs are the words and the groups of bars and dots are the numbers, you were right.

Reading the Numbers

Now take some time to focus on the numbers. How can you figure out what the bars and dots mean? Suppose that one of these symbols counts the 1s. Which do you think it is—the bars or the dots? 🌕

The dots are simple and are similar to the small, round pebbles that people all over the world have used to keep count of things. Let's start by counting them as single marks—1s.

Now look for the largest number of dots that appear together in one row. 🌕

Notice that there are never more than four dots side by side. If the dots are 1s, and there are never more than four dots together, how would you represent five items? 🌕

What about the bars? There are never more than three bars in a group. Could each one be a 5? A 10? Here's a way to explore that question: Assume the bars are 10s. Now try writing the numbers from 1 to 20. (The dots are 1s, and you can use only four dots together.) 🌕

You probably discovered that if a dot is 1 and a bar is 10, there's no way to write the numbers from 5 to 9 or 15 to 19. But if the bars are 5s, you can write all the numbers from 1 to 20.

What Base Are We In?

Now that you can read individual numbers on the codex pages, look for the largest single grouping of bars and dots you can find. (A group is a single row of dots and the bars under it. Some of the numbers may be only dots or only bars.) 🌕

In some places, the number symbols run together, making them hard to read. If you look at the clearest number groups, you'll find that the largest one contains three bars and four dots. If the bars are 5s and the

dots are 1s, what is the value of this largest number represented by this group of bars and dots? Does this give you a hint about the base being used here? 🌕

The number 19—three bars plus four dots—is the largest single number in the Mayan counting system. So how could the Maya write bigger numbers? 🌕

Think about how our number system works. We use ten different symbols, 0 through 9, and combine them to write larger numbers. We can tell what each symbol means from where it appears. For example, "4" means four 1s, but "400" means four 100s, zero 10s, and zero 1s.

The same idea applies to the Mayan counting system. Each group of bars and dots is an individual *digit* that is part of a larger number. The position of the digit tells us what its value is.

If 19 is the largest number the Maya could write in any one position, it makes sense that their system might be base twenty. To represent 20, the Maya would write one dot in the second position, just as we would write a 1 in the second position to represent 10 in our base-ten system. The next position in the Mayan system would be the 400 (20 x 20) place-value position.

Which Way Do We Go?

Now you can begin to figure out what a whole section of numbers in the Mayan codex might mean. Look at how the numbers appear on the pages of the codex. 🌕

You probably notice that most of the groups of dots and bars appear in clear rows and columns.

Archaeologists always choose the clearest areas of information to

A MODERN-DAY MAYAN WOMAN AND HER CHILD IN A MARKETPLACE IN GUATEMALA.

begin their work, so let's begin with the large group of digits (dots and bars) just below the center of the first page of the codex. 🖐

Right away, you face another problem. You can read the individual digits. You even know that they have place values. But how do you know which digits go together to make a number? There are six columns and three rows. Which way do you read? Left to right? Right to left? Top to bottom? Bottom to top? The best way to answer these questions is to choose your best hypotheses and test them. It often helps to hear other people's thoughts when you reach this point, so if you're working with friends, discuss your ideas with them.

If your group is large enough, different people can try different experiments. For instance, one group can look at the symbols as three six-digit numbers read from left to right. Another group can read the same digits from right to left. A third group can read the symbols as six three-digit numbers read from top to bottom. A fourth group can try reading the six three-digit numbers from bottom to top.

Remember, you are working in base twenty, so the first digit is in the 1s place, the second digit is in the 20s place, the third digit is in the 400s place, and so on. Record all the different results, then see if one set of numbers makes more sense than the others. 🖐

If you read the symbols horizontally (either right to left or left to right), you came up with some huge

numbers. After all, the sixth digit in base 20 is the 3,200,000s place! So that's probably not the correct interpretation.

If you read the digits up and down, you came up with more workable numbers. Check to see if you got the following results.

Reading from top 1s to bottom 400s (and left to right):

5,934	4,555	3,535	2,156	1,136	4,117

Reading from bottom 1s to top 400s (and left to right):

5,934	6,151	6,328	6,545	6,722	6,910

Take some time to study these numbers. Can you discover any order or pattern in either of these sets of numbers? 🖐

You might have noticed that the numbers read from bottom to top change in a more regular pattern. They get larger from left to right, and they change at a fairly steady rate. Do you think a set of numbers in a pattern is more likely to carry useful information than a more random set of numbers?

What Does It All Mean?

Now you've figured out that the Maya counted in base twenty. You've also discovered that they wrote and read their numbers from the bottom up, and from left to right. But what were they writing about in this codex? How would you begin to answer that question?

Look at the list of numbers you just deciphered, reading from bottom to top. One way to explore a series of numbers is to find the difference between each pair of numbers:

$$6,151 - 5,934 = 217$$
$$6,328 - 6,151 = 177$$
$$6,545 - 6,328 = 217$$

and so on.

Check to be sure you got these differences:

217   177   217   177   188

What could they mean? The archaeologists who studied the Dresden Codex found an important clue when they added any two of the numbers together. See what happens when you do that.

All of your answers should be between 354 and 434. One of the sums is 365. Does that remind you of anything?

Each of the numbers you added is close to half of 365—the number of days in a year.

An astronomer would also recognize the number 177 as exactly six lunar months of $29\frac{1}{2}$ days each.

It turns out that this part of the Dresden Codex is a record of astronomical observations made by the ancient Maya. This text gives the timing of eclipses of the moon, which occur about every half year.

One More Problem

You may be wondering why all five numbers do not divide exactly into lunar months. To begin to solve the problem, look at the bottom of the same page on the codex. Another easily identifiable group of numbers is there. Use what you have learned to discover what those numbers are.

Check to see if you got these numbers:

177   177   177   177   177   148

Two of these numbers match the differences above, but three do not. This is because Mayan mathematicians used two different numbering systems. For their everyday accounting needs, they used the standard base twenty with its succession of powers: 1, 20, 400, 8,000, and so on, as we have done so far. When working with astronomical calculations, however, they used slightly different bases: 1 and 20 were the same, but instead of 400, they used 360 in the third place value, and 360 x 20—which is 7,200—in the fourth place.

Why? Probably to take advantage of the fact that 360 corresponded more closely to the number of days in the Mayan astronomical year, which was 360 days long, with an additional five days they considered "unlucky." (For more on the Mayan calendar system, see Chapter 7.)

If you want, go back to the groups of symbols near the middle of the left-hand page of the codex. Calculate the symbols again, but this time read the places as 1s, 20s, and 360s.

Check to see if you got these results:

5,374 5,551 5,728 5,905 6,082 6,230

Now find the differences between each successive pair of numbers as you did before.

Check to see that you got these results:

177   177   177   177   177   148

These numbers match the numbers at the bottom of the codex exactly! Five of the numbers are the same: 177. They represent six lunar months of 29 $\frac{1}{2}$ days each. The last number, 148, represents five lunar months (147 $\frac{1}{2}$ days).

This activity was developed by Maurice Bazin and Modesto Tamez.

Making Connections

- Do you know anyone who speaks a language different from the one you speak? If so, ask that person to write something down for you in that language. How different is it from your written language? Is the alphabet the same? Does it have special marks that change the way the words or letters are read? Do you read it from right to left? Left to right? What similarities and differences can you find?

- Some Mayan writing uses glyphs—pictures—that stand for words or parts of words. Can you invent a writing system that does the same thing? What picture would you use for the word "hand"? How about "apple"? Is it harder to create glyphs for words like "sour," "comfortable," or "color"? What could you do to solve this problem?

Recommended Resources

Aveni, Anthony F. *Skywatchers of Ancient Mexico*. Austin: University of Texas Press, 1983.

Coe, Michael D., and Justin Kerr. *The Art of the Maya Scribe*. New York: Harry N. Abrams, 1998.

Coe, Michael D. *Breaking the Maya Code*. New York: Thames & Hudson, 1993.

McLeish, John. *The Story of Numbers: How Mathematics Has Shaped Civilization*. New York: Fawcett Columbine, 1991.

7

THE MAYAN
CALENDAR ROUND

Keeping Time

THE MAYAN CALENDAR ROUND Keeping Time

How often does New Year's Day come around? With the calendar most of us use, a new year begins every 365 days, on January 1. But suppose you had two calendars to follow, each with a different number of days? A new year would begin at a different point on each calendar. How often would the two New Year's Days occur on the same day? How could you figure that out? That's a problem the Maya had.

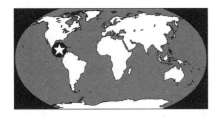

About Mayan Calendars

The Maya, who came to prominence in the New World in the third century A.D., kept extensive records of possessions, important dates, and astronomical observations. They were sophisticated in their understanding of science and math and had an advanced knowledge of astronomy, which they developed through observation and calculation.

For the Maya, religion, math, astronomy, architecture, and culture were all interrelated. People still gather at the Mayan city of Chichén Itzá at the equinoxes (the two days each year when day and night are of equal length) to watch at sunrise, when the shadow of the serpent carved along the grand stairway of the Temple of El Castillo (right) seems to snake its way up the pyramid's side.

The Maya blended together their complex mathematics and precise astronomy in a cycle known as

THE MAYAN TEMPLE OF EL CASTILLO IN CHICHÉN ITZÁ IS A HUGE STONE PYRAMID THAT RISES OUT OF THE MEXICAN JUNGLE ON THE YUCATÁN PENINSULA. IT IS PROBABLY NO ACCIDENT THAT THERE ARE 90 STEPS ON EACH OF THE PYRAMID'S FOUR SIDES, A TOTAL OF 360 STEPS—THE SAME AS THE NUMBER OF REGULAR DAYS IN THE MAYAN CALENDAR YEAR.

the "Calendar Round." They used two different calendars—one for the *haab,* an astronomical year of 365 days, and one for the *tzolkin,* a ritual year of 260 days.

The haab was related to seasonal changes and agriculture, and corresponds to most calendars in use today, with a few differences. It had 18 months of 20 days each. The five days that were left over, known as *uayeb,* were considered unlucky.

The 260-day cycle of the ritual calendar, the tzolkin, can be inferred from the Dresden Codex, one of the few Mayan books that still exist today. (For more about the Maya and the Dresden Codex, see Chapter 6.)

In that manuscript, dates are written as a combination of vertical bars and dots (representing the numbers 1 to 13), followed by one of 20 roundish glyphs, which are

A DETAIL OF THE DRESDEN CODEX, ONE OF THE FEW SURVIVING MAYAN BOOKS. THE BARS AND DOTS REPRESENT NUMBERS (SEE CHAPTER 6); THE ROUNDISH GLYPHS ARE THE NAMES OF INDIVIDUAL DAYS.

WHAT'S IT ALL ABOUT?

In this activity, you'll learn about the two calendars the Maya used. You'll solve the problem of how often the two cycles coincided and find out about:

- What we know about the Mayan calendar system and calendar rounds
- The smallest common multiple of two numbers
- Prime numbers and how they are used to find smallest common multiples

Is there anything special I should know?

- This activity is recommended for ages 12 and up
- You can work alone or with a small group
- You should be familiar with the concept of prime numbers and know how to use a compass and protractor to make circles of different diameters

How much time will I need?

- About 1 to 2 hours

What materials will I need?

- A sheet of posterboard or thin cardboard at least 11 x 14 inches
- A sheet of heavy cardboard at least 9 x 12 inches
- Compass
- Protractor
- Thumbtacks or pushpins
- Scissors

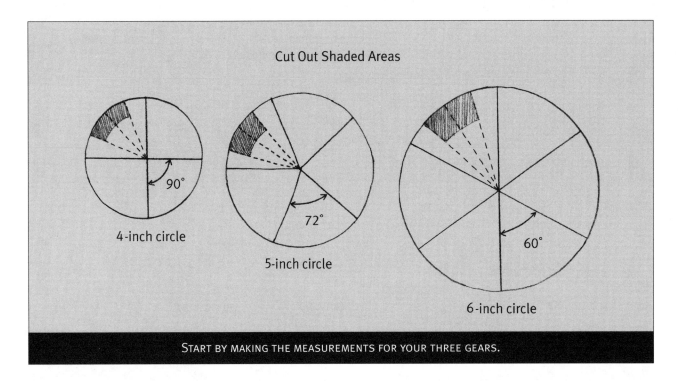

Cut Out Shaded Areas

4-inch circle

90°

5-inch circle

72°

6-inch circle

60°

the names of individual days. Each date is written with one number and one glyph. So the total possible number of days for one cycle of the ritual calendar is 13 x 20, or 260 days.

The Maya believed in the continuing cycles of their world, so the moments when the haab and the tzolkin came together were important events.

Making Gears

In order to figure out how the two Mayan calendars meshed, you need to construct simple gears.

1. Use a compass to draw three circles on the posterboard. One circle should have a 4-inch diameter, one a 5-inch diameter, and one a 6-inch diameter.

2. Use the protractor to draw lines that:

• Divide the 4-inch circle into four equal parts (90° each).

• Divide the 5-inch circle into five equal parts (72° each).

• Divide the 6-inch circle into six equal parts (60° each). (See illustrations.)

3. Each gear needs to have a number of teeth equal to its diameter in inches. That means, for example, that the 4-inch circle will have four teeth. How can you make four teeth and four spaces of equal width? •

• Divide each quadrant of your 4-inch circle into four equal wedges, like slices of pizza. (See the drawing at the top of the page.)

4. Carefully cut out the circles. Then use the lines of the two inner wedges in each quadrant as guides to cut out the teeth. Cut about one-third of the way to the center of the circle.

When you're done, you'll have a gear with four teeth and four spaces of equal width, like the one shown on page 75.

Use the same method to make five equal teeth and spaces on the 5-inch circle, and six equal teeth

and spaces on the 6-inch circle. Again, the drawings will help guide you.

5. Put two of the gears together and see how well they turn against each other. You'll notice that the teeth of one gear probably won't fit perfectly into the spaces of the other. Trim the sharp outer corners of the teeth on all three gears to make sure they mesh together well. (If you look at the gears of a mechanical device such as a clock, you'll notice that the edges of the teeth on those gears are also rounded off to make the teeth and spaces mesh properly.) Trim your gear teeth a little at a time, until they mesh smoothly.

6. When you're done constructing your gears, you need to label them.

• First, label the *teeth* of the 4-inch gear 1, 2, 3, 4, going *counterclockwise*, as shown on the next page. This will represent a four-day year.

• Next, label the *spaces* of the 5-inch gear with the numbers 1, 2, 3,

4, 5, going *clockwise*, to represent a five-day year.

- Finally, label the *spaces* of the 6-inch gear 1, 2, 3, 4, 5, 6, going *clockwise*, to represent a six-day year. ●

Meshing Gears

Each gear you made represents one way of grouping days. In this sense, each one is a calendar.

To begin your investigation, set the 6-inch gear aside. You'll need it later.

Now put a thumbtack or pushpin through the center of the 5-inch gear, and pin it to a large piece of cardboard to hold it in place. Set tooth 1 of the 4-inch gear in space 1 of the 5-inch gear. Then pin the 4-inch gear into place. (See the photo on page 76.) Turn the gears a little to make sure they're meshing well. ●

Now let time flow and the days mesh as you turn both gears. How many turns of each wheel does it take before you're back to the starting position—with the same tooth 1 and space 1 meshing together again? ●

If you let the smaller gear make one revolution clockwise, tooth 1 will be in space 5 of the larger gear. After the next revolution of the smaller gear, tooth 1 will be in space 4. Keep turning the smaller gear until tooth 1 falls again into space 1 of the larger gear. How many revolutions did each gear have to make? ●

You'll find that the 4-inch gear made five revolutions and the 5-inch gear made four revolutions.

To visualize these cycles, imagine putting ink on the edges of the gears and rolling them along a piece of paper in a straight line. One turn of the 4-inch gear gives you a group of four marks. One turn of the 5-inch gear gives you a group of five marks.

How many marks will you have to make before "groups of 4" and "groups of 5" coincide? ●

You'll find that five turns of the 4-inch gear (five groups of 4) will bring you to the same place as four turns of the 5-inch gear (four groups of 5). At the twentieth mark, both will coincide again.

$$4 \times 5 = 20$$
$$5 \times 4 = 20$$

Working with Larger Numbers

If the 4-inch gear and 5-inch gear represent two different calendars of 4 days and 5 days, we can say that there is a 20-day cycle in the system using both calendars. Every 20 days, they will come together again. Once every 20 days, it will be New Year's Day on both cardboard-gear calendars.

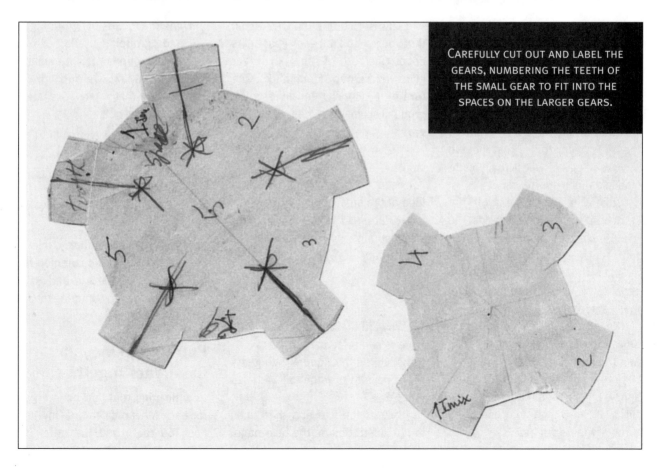

CAREFULLY CUT OUT AND LABEL THE GEARS, NUMBERING THE TEETH OF THE SMALL GEAR TO FIT INTO THE SPACES ON THE LARGER GEARS.

ONCE YOU'VE MADE THE GEARS AND PINNED THEM TO THE BOARD, YOU CAN TURN THEM TOGETHER TO SEE HOW MANY ROTATIONS IT TAKES BEFORE THEY COME BACK TO THEIR STARTING POSITIONS.

But what about the two Mayan calendars, with 365 days and 260 days? Will it take 94,900 days (365 x 260) for the two New Year's Days to happen together again? That's only once every 260 astronomical years!

You can use the 6-inch gear and the 4-inch gear to see that the two New Year's Days will actually coincide in a much shorter time.

Take the 5-inch gear off the cardboard base and set it aside. Pin the 6-inch gear onto the base so that tooth 1 of the 4-inch gear meshes with space 1 of the 6-inch gear. Make sure these two gears mesh smoothly.

Now turn the gears. How many times does each gear revolve before tooth 1 and space 1 line up again? ◐

You'll find that tooth 1 and space 1 mesh again after only two turns of the 6-inch gear and three turns of the 4-inch gear.

Why is this so? Imagine again that you ink and roll each gear on its edge. One turn of the 4-inch gear gives you a group of four marks. One turn of the 6-inch gear gives you a group of six marks. When do the two groupings end together? ◐

The 4-inch and 6-inch gears coincide at the twelfth mark. This corresponds to the mathematical idea of the smallest common multiple—that is, 12 is the first number that contains both "groups of 4" and "groups of 6."

$$3 \times 4 = 12$$
$$2 \times 6 = 12$$

Why did the 4- and 5-inch gears have to complete the full 20 (4 x 5) turns while the 4- and 6-inch gears didn't have to turn the full 24 times (4 x 6)? ◐

Look at the first and third lines in the illustration on the next page,

showing the results of "grouping by 4s" and "grouping by 6s." Notice that each group contains a smaller internal grouping—"groups of 2." That means that 4 and 6 share a common factor of 2.

$$6 = 3 \times 2 \text{ (three groups of 2)}$$
$$4 = 2 \times 2 \text{ (two groups of 2)}$$

If you multiply 6 x 4, then factor out the common 2 (divide by 2), you get the smallest common multiple of 6 and 4, which is 12.

There is no smaller common internal grouping—no common factor—for the numbers 4 and 5, so those gears had to complete the full 20 turns.

Putting the Mayan Calendars Together

Now imagine that you have a gear for each Mayan calendar. One gear has 260 teeth and the other has

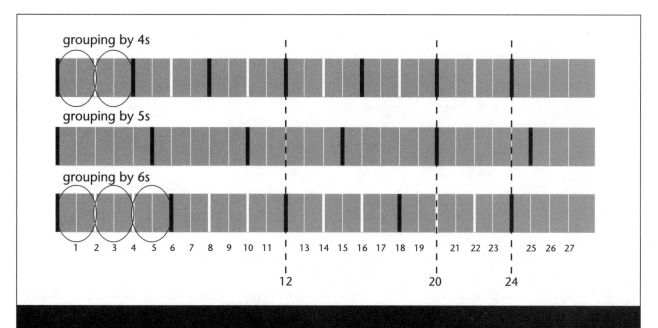

grouping by 4s

grouping by 5s

grouping by 6s

1 2 3 4 5 6 7 8 9 10 11 12 13 14 15 16 17 18 19 20 21 22 23 24 25 26 27

NOTICE THAT WHEN YOU COUNT BY 4S AND 5S, THE FIRST PLACE THEY COINCIDE IS AT THE 20 MARK (4 x 5). BUT THE 4S AND 6S COINCIDE AT BOTH THE 12 MARK AND THE 24 MARK, BECAUSE 4 AND 6 SHARE A COMMON FACTOR OF 2.

365 teeth. When we mesh tooth 1 of one calendar with space 1 of the other, it's the first day of the year for both calendars. Then time moves on and the gears turn. How long will it take before the two years begin on the same day again?

To find out how long it will take for the two Mayan calendars to coincide, you'll need to find the smallest common multiple for the numbers 365 and 260. To do that, you need to start by finding all the numbers by which 365 and 260 can be divided exactly.

This process is called finding the prime factors of a number. Prime factors are all the factors of a number that are prime numbers.

Prime numbers are numbers that can be divided exactly only by themselves and 1. For example, 7 is a prime number because it can be divided only by 1 and itself. But 9 is not a prime number because it can also be divided by 3.

The first ten prime numbers are 1, 2, 3, 5, 7, 11, 13, 17, 19, and 23.

To find the prime factors of 260, the number of days in the Mayan ritual calendar, begin by dividing 260 by 2. Keep dividing by 2 until the result is an odd number. (You can't divide an odd number exactly by 2.) Then try dividing that odd number by the next prime number—3. Divide by 3 as many times as you can. Then try dividing by 5, and so on.

Try finding the prime factors of 260. 🌑

The prime factors of 260 are 2 x 2 x 5 x 13.

Now do the same thing with the prime factors of 365, the number of days in the Mayan astronomical calendar. 🌑

You probably discovered quickly that you can't divide 365 exactly by 2 or by 3. But you can divide it by 5, which gives you a result of 73. When you try to divide 73 by successive prime numbers, you will discover that 73 is itself a prime number!

The number 365 can only be factored into 5 x 73. The numbers 260

and 365 share only one common factor: 5.

$$260 = 5 \times 52$$
$$365 = 5 \times 73$$

Now that you know this, can you figure out the smallest common multiple for 260 and 365? 🌑

If you factor out the common 5 (that is, divide by 5), you get 18,980 (5 x 73 x 52), which is the smallest common multiple of 260 and 365.

What does that tell you about when the two Mayan calendars coincide?

As you have discovered, exactly 18,980 days will pass before the two New Year's Days occur on the same day again. They will come together again after 52 astronomical years of 365 days each, which is the same amount of time as 73 ritual years of 260 days each.

The Mayan Calendar Round

The 52-year cycle in Mayan life has been called a "calendar round"

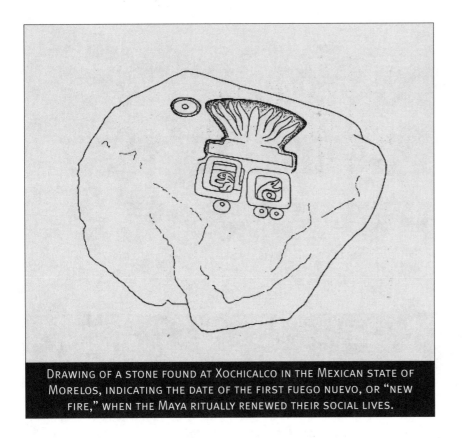

DRAWING OF A STONE FOUND AT XOCHICALCO IN THE MEXICAN STATE OF MORELOS, INDICATING THE DATE OF THE FIRST FUEGO NUEVO, OR "NEW FIRE," WHEN THE MAYA RITUALLY RENEWED THEIR SOCIAL LIVES.

because it involves the ways of counting days in both calendars, and leads to a new start for both calendars at the same time. Scholars have found a large numerical table of multiples of 52 in the Dresden Codex.

Anthropologists have known for a long time that a cycle of 52 astronomical years was important in the social life of the Maya. Every 52 years the Maya extinguished all fires in their households and threw away all their clay utensils. Then they "renewed" their social life by bringing a new fire (*el fuego nuevo*) from a central location into all the villages and cities.

The last time the Maya celebrated el fuego nuevo, the beginning of the 52-year cycle, was shortly before the Spanish conquest. By our calendar, that was in December of 1507. For the Maya, this last celebration of the calendar round occurred in the month of *Yaxkin*.

Counting for a Long, Long Time

The Maya had an even longer view of their calendar than the cycle of the haab and tzolkin together. There is a calendar called the "Long Count" that was used to record dates over long periods of time. The long count "year," called *a tun*, has 360 days. A period of 20 tuns (7,200 days) is called a *katun*, and 20 katuns (144,000 days) is called a *baktun*.

A period of 13 baktuns is called the "Great Cycle" of the Long Count. It is 5,200 years long (in 360-day years). The current Great Cycle will come to an end on what most of us will call December 21, 2012 (the winter solstice), which will be a significant date to modern Maya people.

This activity was developed by Maurice Bazin and Modesto Tamez.

Making Connections

• Ask your grandparents, or the oldest people you know in your family or community, about times of renewal, such as New Year's Day, Chinese New Year, and Rosh Hashanah. Do they know of other traditions that were practiced before you were born or in other places?

• Your birthday probably comes once a year. But what if you were born on February 29? That date occurs only every four years, during a "Leap Year." How often could you celebrate your real birthday? How "old" would you be now if you counted only that day as your birth date? How "old" would an elder of your family or community be?

Recommended Resources

Coe, Michael D. *The Maya*. New York: Thames and Hudson, 1999.

Malmstrom, Vincent H. *Cycles of the Sun, Mysteries of the Moon: The Calendar in Mesoamerican Civilization*. Austin: University of Texas Press, 1997.

Sharer, Robert J. *Daily Life in Maya Civilization*. Westport, CT: Greenwood Publishing Group, 1996.

Social and Cultural Traditions

8

WEAVING BASKETS

Discovering Patterns and Symmetries

WEAVING BASKETS
Discovering Patterns and Symmetries

For many thousands of years, people all over the world have made baskets. There are as many materials and techniques for weaving baskets as there are communities that make them, but all basket weaving follows similar mathematical patterns.

The Ancient Craft of Basket Weaving

Weaving is one of the oldest crafts in the world. Imprints of woven baskets and clothing have been found in sites more than forty thousand years old.

Over the centuries, different cultures have woven their baskets using different materials—from grasses and pine needles to strips of wood and paper. As with other handcrafted items, a culture's traditional baskets are made of local materials. Native Americans have made baskets using a wide variety of plants and plant materials, including bark, grass, corn husks, vines, and ferns. They have even woven baskets from horsehair and whale baleen.

European settlers in North America often wove baskets from thin strips of hardwood, such as ash, oak, and hickory. Contemporary American basket makers still use local plants and hardwoods, as well as reed and cane imported from Asia.

Because baskets are used for so many things, there is a tremendous variation in how they are designed.

WHAT'S IT ALL ABOUT?

In this activity, you'll weave flat paper mats and then start to notice some of the mathematical patterns in their structure. You'll also have a chance to investigate:

- Two different weaving techniques: plain weave and twill
- What happens when you experiment with the pattern of a weave
- What symmetry is and where it is found in the patterns of a woven basket

Is there anything special I should know?

- This activity is recommended for ages 10 and up
- You can work by yourself, or in a small group

How much time will I need?

- Allow about 2 1/2 hours to do the weaving and look at the patterns

What materials will I need?

- A variety of different baskets to examine
- Two different colors of heavy paper, cut into strips about 12" long and 1/4" wide; each person should have 40–60 strips of each color (NOTE: If you have access to a paper shredder, feed used file folders into it to get lots of perfect 1/4" strips)
- Masking tape
- Pencils or colored pencils
- Handheld mirror with one straight edge
- Graph paper (optional)
- Tracing paper (optional)

Woven baskets have been used to store food, clothing, and utensils. They have been used for clothing (from hats to shoes), in transportation (from boats to the baskets that hang under hot air balloons), and even for trapping and fishing.

Some baskets are woven to be loose and open; others are compact, with strands placed as close together as possible, like finely woven fabric. Some are so tightly woven that they even hold water.

Most baskets are made to be both utilitarian and beautiful. Even when creating a basket for a specific job, weavers often go to a great deal of trouble to make the basket pleasing to the eye.

Most baskets are woven in symmetrical patterns. Some have elaborate designs woven in or are made from more than one type of material. Native Americans sometimes use beads, quills, and feathers in their weaving. Dyes may also add color and create patterns in the weaving.

In some areas of the world, the art of basket making has begun to disappear. The need for functional baskets has declined, and people have chosen other ways to make their living. But there are also places where basket makers are working to keep their craft alive. Traditional basket makers in Scotland use willow that they grow and harvest. Nantucket baskets, traditionally made on the island off the coast of Massachusetts, have become popular and valuable. And Native American baskets, old and new, are highly valued for their beauty and history. A renewed interest in baskets has allowed traditional artisans around the world to keep this ancient craft alive.

Looking at Baskets

Before you begin your own weaving, take time to look carefully at some

baskets and see what you can learn about how they were made. The patterns and weaves may range from simple to complex. Some may have designs with different colors woven in. As you look at each one, think about these questions:

- What material(s) is the basket made of?

- Was the basket woven with a single pattern or several different patterns?

- Does the bottom of the basket look different from the sides?

- What do you think each basket might be used for?

Making a Plain-Weave Mat

There are many different weaving techniques, but one of the most basic is called *plain weave*. Plain weave uses two groups of strips that are woven perpendicular to each other.

When you start your own weaving, you may want to use strips of two different colors—one color for the vertical strips, and another color for the horizontal strips. The vertical strips are called the *warp*. The horizontal strips are called the *weft*.

Follow these directions to make

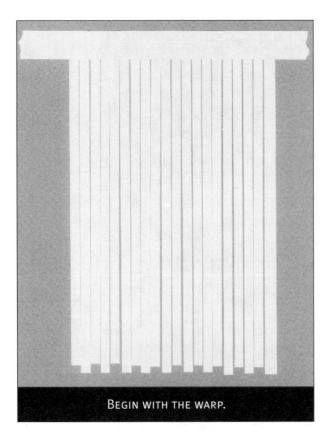

BEGIN WITH THE WARP.

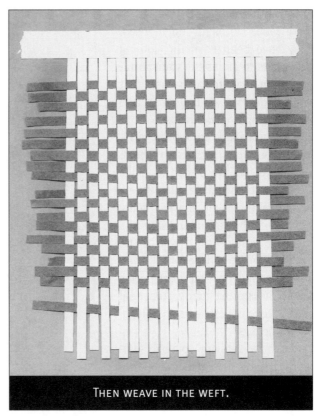

THEN WEAVE IN THE WEFT.

a plain-weave mat. The photos above will help guide you.

1. Choose 20 strips of one color for your warp. Lay them vertically on a table and push the strips up close to each other. Hint: If you let the bottom ends of the warp hang off the edge of a table a bit, you'll be able to pick them up easily. It also helps if you arrange the strips so that every other one sticks out a little.

2. Once you've arranged the warp strips, put a piece of masking tape across their tops to hold them in place.

3. Now you're ready to start weaving. Choose a strip of paper in the second color to start the weft. Weave the strip horizontally through the warp, going under the first strip, over the second, under the third strip, over the fourth, and so on. Keep going until the horizontal piece is

woven completely through all the vertical strips.

4. Push this horizontal weft strip up near the top of the warp and choose a new strip of the second color. Weave it into the warp just like the first one, except this time start by going over the first strip (instead of under it), under the second strip (instead of over it), and so on. When the second strip is completely woven in, push it up against the first weft strip. Weave the third weft strip exactly like the first one and the fourth strip exactly like the second one.

5. Continue in this way, weaving each strip in the opposite over/under pattern from the strip above it. Push each new weft strip firmly up against the previous one. You may need to push the warp strips together too, to make the mat tight and firm.

6. When you've used 20 warp strips and 20 weft strips, you should have a mat of plain weave that looks like the one pictured above.

Pick up your mat and take a look at your weaving. Is the pattern the same horizontally and vertically? Is it the same on the front and the back? Is the weave strong and regular? Are all the horizontal strips fully woven into the pattern? ✋

Now set your plain-weave mat aside and try a different weaving technique.

Weaving in Twill

Now that you know how to do plain weave, you can try another kind of weave called *twill*.

The structure of twill is more complex than plain weave. Twill creates a visual pattern of diagonal lines from the horizontal and vertical strips. There are many twill

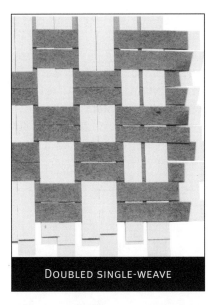

DOUBLED SINGLE-WEAVE

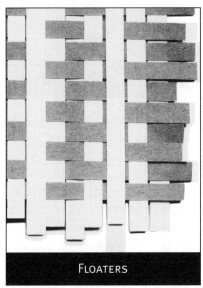

FLOATERS

TWILL WEAVE

designs, each made with a different over/under pattern. Instead of the unique one over/one under of plain weave, a twill weave might be one over/two under, or three over/two under.

In this activity, you'll weave a twill with a pattern of two over/two under. To begin, you can start the way you started the plain weave.

1. Make a vertical warp of about 20 strips. Remember to let the strips hang off the edge of the table a bit. You should still tape the tops of the strips to the table with masking tape, but there's no need to stagger every other strip.

2. Once the warp is ready, you can start weaving the horizontal strips in a two over/two under pattern. 🌀

    Weaving in the first strip should be easy. But you might wonder where to start weaving in the second weft strip. This is a good time to do some experimentation, as basket makers throughout history have done.

    A proper weave has definite rules. For the plain weave, the rule was one over/one under, alternating every other strip. What sort of rule could you try for your twill pattern? Test it by weaving a few strips. 🌀

    If your rule worked, you can continue weaving. If it didn't—if you ran into a place where there were loose strips or irregular gaps in the weave—you'll need to start over using a new rule. You want to find a way to make a mat with a twill weave that doesn't gap or split apart.

3. When you find a pattern you like, weave in about 20 horizontal strips. 🌀

Testing the Weave

When people try to figure out how to make a twill mat, they often start by weaving in the horizontal strips two at a time. This gives a two over/two under weave, but it doesn't create visual diagonals. In fact, this is a version of plain weave that simply uses doubled strips. In a large piece woven in this pattern, the two side-by-side strips can slip around, and end up overlapping one another. As a result, this kind of weaving gives the mat less structural strength than plain weave itself.

You may have tried staggering the weft strips, skipping one warp strip when you wove in the second horizontal strip in a two over/two under pattern. If you weave the third strip like the first, and the fourth strip like the second, you'll get an odd surprise. Every fourth vertical strip becomes a "floater"— a strip that is not woven into the mat. If you turn the mat over, you'll see more free-floating strips.

The only successful two-by-two twill weave is made by shifting each horizontal strip over, one strip at a time, and always in the same direction. This systematic process creates a distinctive diagonal pattern and also keeps the strips locked together without shifting or loosening up. You'll see a visual diagonal pattern emerge at a 45-degree angle to the strips of warp and weft.

You can see this kind of twill in the Brazilian basket on page 85, and in the paper mat in the photo at the bottom of this page. Look closely to see the diagonals in the patterns on the sides of the basket: The diagonals run parallel to the base.

Now pick up your mat and take a look at your weaving. Is the pattern the same horizontally as it is vertically? Is it the same on the front as it is on the back? Is the weave strong and regular? Are all the strips fully woven into the pattern? ◐

Recording the Patterns

As you practice these two different weaving techniques, you might find it useful to draw them. Not only does this let you keep a record of your experiments, but your drawings may also reveal structures you didn't notice as you were weaving.

There are many ways to record the structures in your weaving. Experiment with a pencil and some plain paper or graph paper. Take a good look at one of the mats you've woven. One thing you need to decide is how to make a two-dimensional drawing represent a three-dimensional weaving. How can you show which strips go over and which strips go under? ◐

One way to do this is to draw each segment as a short line, with a break where the strip went under another strip. Another way to record your pattern is to use two different colors to represent the warp and weft of your weaving. Using graph paper may help you organize the horizontal and vertical lines. ◐

If you're having trouble drawing the structure of your mat, you might try tracing one of the weavings you've just made. Put a piece of tracing paper over your plain-weave mat or your twill mat and use a pencil to trace each strip's path. ◐

Whatever method you use, once you've made a drawing, the structure of your mat will probably be more clear to you. Here are two examples, one showing the structure of a plain-weave mat, the other the structure of a twill mat.

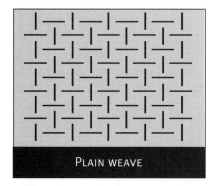

PLAIN WEAVE

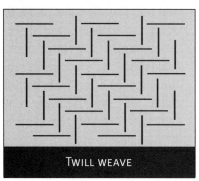

TWILL WEAVE

Looking for the Pattern's Core

Pick up your plain-weave mat and take a careful look at the structure of the over/under pattern. Turn the mat 90 degrees so that the horizontal strips are now vertical. Is the over/under pattern the same as before? Now flip the mat over and see if it is the same on the back as it was on the front. ◐

You'll notice that no matter which way you turn the plain-weave mat, its structure looks the same.

Weaving is a repetitive activity. The overall pattern of your plain-weave mat is actually a repetition of a much smaller pattern: two horizontal strips interwoven with two vertical strips, forming four small squares, like a tiny checkerboard. You could make a huge mat of plain weave by just repeating that pattern of four squares over and over. The pattern of the four squares is called a *minimum block*.

If you draw a 20 x 20 mat, coloring in the squares on graph paper,

you can see the repetition clearly. The minimum block is the smallest pattern you can find before the structure's design starts repeating itself. If you shifted, or "translated," one minimum block sideways, it would match the next block exactly. If you shifted it diagonally, it would match the next block again.

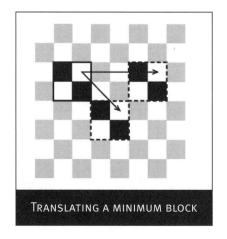

TRANSLATING A MINIMUM BLOCK

But what about twill? Try to figure out the minimum block for twill by taking a close look at the twill mat you made. You want to find the smallest structure in the weave that repeats itself over and over to make up the whole mat. ◐

When you look closely at your twill mat, you'll find that the minimum block is four strips by four strips. Since the fifth strip in each direction is woven in exactly the same way as the first strip, it starts a new minimum block.

TWILL MINIMUM BLOCK

Finding the Symmetries in Plain Weave

When you look at your woven mats, you might be struck by the intricate patterns in the weaves. These patterns make baskets both attractive and structurally sound.

The decorative shapes and designs of the woven patterns are often quite regular: They may form squares or crosses or zigzags. One way of looking at these patterns is to examine the mathematics of their structure.

Some regular patterns are symmetrical, which means that they repeat in specific ways. There are many different kinds of symmetry. When you shifted a minimum block of plain weave or twill diagonally, or to one side, and discovered that it matched the next block exactly, you were looking at *translational symmetry.*

To see other kinds of symmetry, look at a single minimum block in your plain-weave mat. Turn the mat 90 degrees, so the horizontal strips are now vertical. Then turn it another 90 degrees, so it's upside down from where it started. Is the pattern different? ◐

The basic pattern of plain weave will be the same no matter which way you turn it, and no matter how many 90-degree turns you make. This kind of symmetry is called *90-degree rotational symmetry,* or *fourfold symmetry.*

Your plain-weave mat is symmetrical in another way as well. Put your mat on a table and put the edge of a mirror along a strip that goes through the middle of the mat. (See the illustrations on page 90.) You can see one half of the mat on the table and one half of the mat in the mirror. What do you notice? ◐

How does a basket maker turn a flat mat into a three-dimensional basket? Here are the first few steps in the process:

1. START WITH A PLAIN-WEAVE MAT WITH BOTH WARP AND WEFT ENDS LOOSE. (THE BLACK STRIP AT THE TOP OF THE MAT IN THE PHOTO IS A REFERENCE POINT, SO YOU CAN SEE WHAT'S HAPPENING AS YOU WEAVE.)

2. BEND THE MAT SLIGHTLY IN THE MIDDLE. START TO WEAVE TOGETHER THE FREE STRIPS FROM THE TOP OF THE MAT. BEND THEM TOWARD EACH OTHER AND USE AS MANY STRIPS FROM THE LEFT AS FROM THE RIGHT.

3. THIS IS THE INSIDE OF THE BOTTOM OF THE BASKET. THE STRIPS NOW MEET AT 90 DEGREES, FORMING A CORNER. (IT IS EASIEST TO SEE THIS IF YOU FOLLOW THE BLACK STRIP, WHICH IS NOW FOLDED AT 90 DEGREES AND IS PART OF THE BASE AND ONE OF THE SIDES.)

If you look in the mirror, the reflected half of the mat looks identical to the half of the mat behind the mirror. This is called *mirror symmetry*. If a pattern has mirror symmetry, you can fold it in half and the two halves will match up exactly.

What happens if you place the mirror's edge diagonally across the center of the mat?

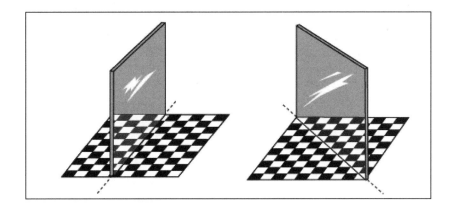

Balaio

Discovering the Mathematics in Basket Making

While in the Alentejo region of Portugal, I visited a basket weaver who wove slats prepared from young chestnut trees. I was interested in the long, complicated process used to prepare the wood, but another aspect of the weaver's work attracted my attention. After weaving a base and bringing up the free ends on all four sides to make the ribs of the walls, the weaver added one extra side rib. He did this, he said, to "guarantee that the wall winds around properly."

This comment referred to weaving a long, continuous strip of material around the side ribs of the basket so they would wind alternately in and out of the ribs on each successive turn. The weaver expressed his mathematical translation of this structural requirement in this way: "You need an odd number, like five, seven, or nine; an even number will not do."

On another occasion, friends in Rio de Janeiro told me about the large baskets their father often made to carry coffee on plantations. We visited this man, who cut down some bamboo stalks and made us one of these carrying baskets, called a *balaio*. Even though the balaio was built with bamboo, it began to look

very much like the basket I'd seen in Portugal.

I waited for the weaver to stick an extra rib into the side. Instead, he slashed one side of one of the cross ribs, splitting it into two ribs. That was his way of creating an odd number of ribs. When I asked why he did it that way, he said that as far as he knew, that was the way it had always been done. The "why" had become what mathematician Paulus Gerdes calls "submerged mathematics." The weaver may not have known the mathematical reasons for adding a rib, but the mathematical requirements were there, both in Portugal and in Brazil.

When you pick up baskets here and there, look at the number of ribs around the sides of each one. You may be surprised, as I was, to find an "impossibility"—a basket made with an even number of ribs. How could that be? When I examined one of these baskets closely, I saw that each horizontal strip had been cut after one full turn so that each strip was independent of the others. That basket was clearly asking to unwind—it lacked the strength it would have had if a single horizontal strip had been woven continuously around an odd number of ribs.

—Maurice Bazin

Again, the mirror image matches exactly the part of the mat you can't see. The entire pattern is completed again.

Looking for Symmetries in Twill

Now you can look for symmetries in your twill mat (or the drawing you made of it). Try turning it, folding it, and looking at it with a mirror in different positions. Can you reproduce the original pattern?

As strange as it may seem, the twill pattern does not display any symmetry other than the translational symmetry of its minimum blocks. This puts limits on the basket weaver's creativity. If you look at the twill baskets shown on this page, you'll notice that the regular twill pattern is broken in several places in the base of each basket. This is because you cannot make a symmetrical basket (like one with a square base) when you start from a basic pattern that has no symmetry. Since the twill pattern is not symmetrical, all twill baskets will show these breaks. How and where to make them is a challenge to each weaver's creativity.

Revisiting Your Baskets

Now that you know more about weaving and basket making, look again at the baskets you collected. Can you figure out how they were woven? Do they display any symmetries you now recognize? If you know how to look, you may be able to find patterns and symmetry all around you.

THE PAPER BASKET (TOP) WAS MADE FROM A TWO-COLOR, 12-BY-12 STRIP MAT. THE CENTER BASKET WAS WOVEN BY A GUARANI FAMILY OF PALHOÇA, SANTA CATARINA, BRAZIL. THE RIMMED BASKET (BELOW) IS FROM THE PHILIPPINES. THE SIDES OF EACH BASKET ARE ENTIRELY TWILLED, BUT EACH BOTTOM MUST BREAK THE TWILL PATTERN TO CREATE A FOURFOLD SYMMETRY AROUND ITS CENTER.

This activity was developed by Maurice Bazin and Kim Shuck.

Making Connections

- Baskets have been made all over the world for thousands of years. Why do you think this art form is so widespread?

- How does a basket's use affect the way it's made? How it looks?

- How are baskets used where you live? Are there special times or occasions when you use them?

- What materials could you find in your area to weave baskets? What modern materials are used to weave baskets today?

Recommended Resources

Hoppe, Flo. *Contemporary Wicker Basketry.* Asheville, NC: Lark Books, 1997.

La Plantz, Shereen. *Twill Basketry.* Asheville, NC: Lark Books, 1993.

Philippoff, Jennifer, and Diane V. Maurer-Mathison. *Paper Art: The Complete Guide to Papercraft Techniques.* New York: Watson-Guptill Publications, 1997.

9

DYEING

Colors from Nature

DYEING Colors from Nature

For thousands of years, people have used natural dyes to create colorful clothes and fabrics. In this activity, you'll dye your own yarn using natural materials.

A History of Dye Colors

The history of dyes is closely connected to social history. The harder it was to get a particular dye, the more expensive it was. This also gave dyes a symbolic social value—only the rich and powerful could afford clothing dyed with expensive colors.

The most common natural dye colors are yellow and a variety of greens. Dyes that can produce colors at the ends of the spectrum, such as reds, blues, and purples, are much more rare. Since these unusual colors contrasted with the earth tones that most people wore, they were the preferred colors for royalty and military rulers.

In ancient Rome, for example, only the emperor and a few high officials were allowed to wear purple robes. Red trousers distinguished the royal troops in Europe. Laws that governed the colors and the fabrics used for clothing—called *sumptuary laws*—helped define class distinctions. The colors and fabrics people wore were symbols of their position in society.

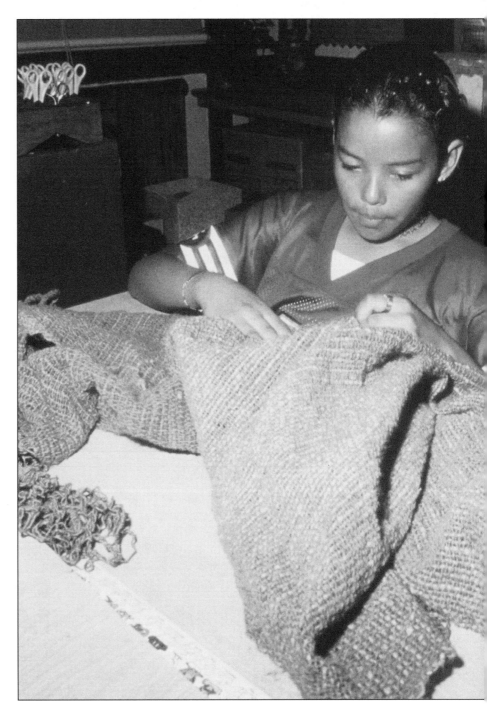

MADDER'S RED COLOR COMES FROM TWO COMPOUNDS, PURPURIN AND ALIZARIN.

Purpurin
(from madder root)

Alizarin

Making Reds

Red dye can be made from the sweet woodruff plant or from the root of the madder plant (*Rubia tinctorum*).

Red dye made from madder probably had at least two separate origins. It was used by the Vikings who settled at York in Great Britain. It was also used in India, transported to Turkey and the Middle East, and then taken west by the French, who called it "Turkey red" because it was the color of the traditional Turkish hat, the fez. Madder was later imported by Great Britain and used to dye the uniforms of the "Redcoats" during the American Revolution.

In Europe, there was great competition for control of madder cultivation. Holland grew the most madder in the seventeenth century; France took over in the middle of the eighteenth century. In 1869, the competition came to an abrupt end when the first synthetic dye was made in England. It was cheaper to produce, but thousands of European farmers who had been growing madder were put out of work.

Another red dye, called *cochineal*, comes from small, bright red insects (*Dactylopius coccus*) that live on cactus plants in central Mexico. Until the sixteenth century, Aztecs collected the female insects and boiled them to dissolve the red dye molecules out of their bodies.

Making Blues and Purples

At the other end of the spectrum are the blues and purples. Blue dye comes mainly from two sources, the woad plant (*Isatis tinctoria*) from Europe and the indigo plant (*Indigofera tinctorum*) from India.

Woad, a member of the mustard family, is well adapted to cold climates. Its chemical content is identical to that of indigo dye (*indigotin*), but less concentrated. Immediately after being dyed with indigo, fibers turn green. The green quickly turns to blue as the dye oxidizes.

The Celts, the infamous "blue barbarians," smeared themselves with woad before charging into battle. Druids used woad to tattoo

themselves in serpentine patterns. Well into the eighteenth century, European countries passed laws forbidding manufacturers to import blue indigo dye from India, in order to protect local woad growers.

Indigo originally grew in India and Egypt. Portuguese navigators brought it to Europe in the sixteenth century. In the mid-seventeenth century, British-American settlers in Virginia planted indigo, thinking it would be even more profitable than tobacco and would allow them to take over India's trade. It took one hundred years before indigo was successfully grown in the United States, and then only in South Carolina, which shipped one million pounds of indigo yearly to England in the 1700s.

The uniforms of the American troops during the Revolutionary War were dyed with indigo. Today, indigo is used to dye cotton denim to make

Carminic Acid
(from cochineal)

THE COLOR-PRODUCING COMPOUND IN COCHINEAL IS CARMINIC ACID. YOU MAY NOTICE THAT THE STRUCTURES OF PURPURIN, ALIZARIN, AND CARMINIC ACID LOOK SIMILAR: THEY EACH CONTAIN THREE RINGS OF CARBON ATOMS. VIBRATIONS OF THE ELECTRONS IN THESE RINGS ABSORB THE BLUE END OF THE LIGHT SPECTRUM, LETTING US SEE THE RED COLOR.

blue jeans. It is a very popular dye because it works well on natural fibers like wool and cotton and does not fade in light or water.

In the ancient world, the color purple—the rarest natural dye of all—was reserved for the most exclusive segments of society. Purple robes were worn only by the noble classes of ancient Greece and Rome. In the late fourth century, Emperor Theodosium of Byzantium decreed that certain shades of purple could be worn only by the Imperial family—upon pain of death.

The rich hues of royal purple were harvested from the glands of two long, spiral-shelled mollusks, murex

WHAT'S IT ALL ABOUT?

This activity is divided in two steps to be done over two days. First, you'll prepare the wool using a process called *mordanting*. Mordanting makes the color brighter and more permanent. It should be done the night before the dyeing. In the second step you'll dye wool (yarn) that can later be used to make clothing, a rug, a blanket, and so on. In the process, you'll have the opportunity to explore:

- Why dyes were originally developed
- How dyes of different colors are made
- What causes a dye to adhere to cloth
- What you can do to change the color you get from a particular dye

Is there anything special I should know?

- This activity is recommended for ages 10 and up
- You can do this by yourself or in a small group
- If children are doing this activity, an adult should supervise the process
- Note that alum is light-sensitive, so be sure to cover the pot during the mordanting process
- Though the chemicals used in this activity have low toxicity, it is always important to wear eye protection when working with any chemicals ⚠
- This activity uses boiling water, which can be dangerous and cause serious burns ⚠

How much time will I need?

- Each step will take about 1½ hours, but note that the mordanted wool needs to soak overnight before being dyed. The eucalyptus leaves may also need to be soaked. See the note below.

What materials will I need?

- Eye protection ⚠
- Measuring cups and spoons
- Source of water
- Stovetop for boiling water
- 10 grams (about 2¾ teaspoons) alum (available in the spice section of most supermarkets)
- 4 grams (about 1½ teaspoons) cream of tartar (also available in supermarket spice section)
- Large nonreactive pot (stainless steel, Pyrex, or porcelain), with a lid
- 50 grams (about 2 ounces) of raw or natural unbleached wool yarn, found in knitting supply stores

- A stick or long-handled wooden spoon (to poke the yarn down into the dye)
- Nonmetal colander
- Dark-colored plastic garbage bag
- Plastic or wooden slotted spoon (optional)
- One or two plastic hangers
- Plastic bucket or bowl (if you're using dried eucalyptus leaves, see note below)
- 100 grams (about 4 ounces) of eucalyptus leaves, dried or fresh (in some places, notably California, these can be obtained in the wild; they are also available at florists and botanical gardens; if eucalyptus is not available, yellow onion skins may be substituted)

NOTE: We found that dried eucalyptus leaves produced the most even color distribution. However, if you use them, they'll need to soak overnight (the same night as the wool). Just crumble them into a plastic bucket or bowl, cover with water, and let them sit until you're ready to make your dye. If you're using fresh eucalyptus leaves, you don't have to prepare them in advance.

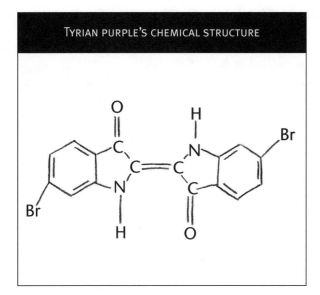

TYRIAN PURPLE'S CHEMICAL STRUCTURE

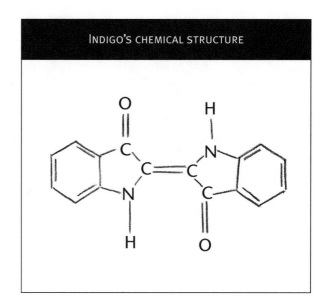

INDIGO'S CHEMICAL STRUCTURE

(*Murex trunculus*) and purpura (*Purpura haemastoma*), which are found along the eastern shores of the Mediterranean. The color was known as Tyrian purple because the Phoenicians processed it in the city of Tyre.

Artificial Dyes

Mauve—a sort of purplish-pink color—was the first artificial dye. It was originally produced in 1856 by William Henry Perkin, a student at the British Royal College of Chemistry. Perkin was trying to synthesize quinine from aniline sulfate. Instead, he created a violet-colored substance that he called *mauvine*. Though he was only eighteen, Perkin recognized the potential of his find, obtained a patent, and built a factory to supply mauvine to the London silk dyers.

Mauvine was a great source of wealth and commercial power, and other synthetic dyes soon followed. Perkin saw England become the greatest color-producing country in the world. England exported tar-distilled crimsons to cochineal-producing Mexico, and coal-derived blues to indigo-growing India. While

this commerce helped England's economy, it contributed to the ruin of India's indigo-based textile industry.

Dyeing Your Own Wool

Since your yarn will need to be prepared for the dyeing process, this activity has been designed to be done on two consecutive days.

Day 1: Preparing the Wool

Mordanting is a process of treating wool by boiling it in water with metal salts. The word "mordanting" comes from the Latin word *mordere*, meaning "to bite." The mordant provides a metal ion that "bites" into each molecule, wool and dye, forming a bridge between them.

Some natural dyes do not need additional mordants. For example, eucalyptus bark and onion skins contain enough tannins to dye wool that hasn't been mordanted. However, mordanting with these dyes gives a wider range of bolder colors and makes the dyes more colorfast.

Alum is hydrated potassium aluminum sulfate. It is used for making

pickles (that's why it's sold in the supermarket) and is a favorite mordant because it is the least irritating to the skin.

1. Dissolve the alum and cream of tartar in about 100 milliliters (ml) of water (about 3$\frac{1}{2}$ ounces). Then put the mixture in the large, nonreactive pot filled with 2 liters (about 2 quarts) of water.

2. Put the clean, damp wool into the pot and bring it to a boil. Turn down the heat and let it simmer for ten minutes to an hour. Alum is light-sensitive, so try to keep the lid on the pot.

3. Let the pot cool and leave the wool to soak overnight, still covered in the pot. If you're using dried eucalyptus leaves, they'll also need to be soaked overnight. See the note in the materials list.

Day 2: Dyeing the Wool

1. Take the wool out of the pot and drain it in the nonreactive colander.

2. After a few minutes, take the wet wool out of the colander and put it into the dark-colored plastic bag. Seal the bag and set it aside. You'll need the wool again in a little

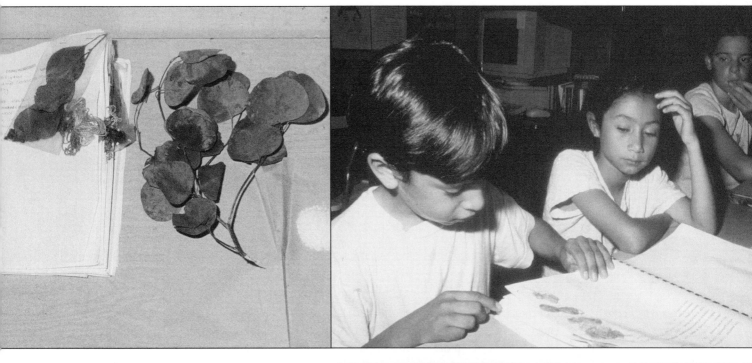

while, but for now you'll, want to keep it out of the light.

3. Rinse out the nonreactive pot and put the eucalyptus leaves (fresh, or dried and soaked, as noted in the materials section) into the pot. Add enough water to completely cover the leaves. Then set the pot boiling to make the dye.

4. Simmer the dye for 30 minutes to an hour. The amount of boiling time depends on how thick the leaves are and what color dye you want. Feel free to experiment.

5. After the dye has simmered, let it cool down to lukewarm. If you want, use a slotted spoon to remove the leaf bits from the dye bath. (It doesn't hurt anything to leave them in, though.)

6. Add the damp, mordanted wool to the dye. Stir gently with a stick or wooden spoon to make sure all the wool is covered by the dye. (Be sure you don't press too hard, though, or the wool will *felt*, or mat together.)

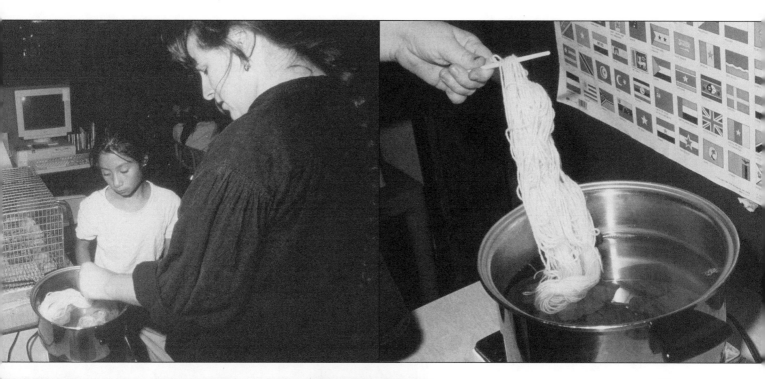

7. Let the wool sit in the dye for 30 minutes or more. Try varying the amount of time the wool is in the dye, and see what happens.

8. Remove the wool from the dye and drain it in the colander for a few minutes. Then drape it over a plastic hanger and let it dry. (It will drip, so put it over the sink or a bucket.)

The resulting color may range from tan to peach, depending on the species of eucalyptus used, the amount of time the wool is left in the dye bath, and whether or not the wool was mordanted.

The Chemistry of Dyeing

The most common use of dye is to color fabric. The colors you get result from the physical properties of the dyes that you use.

All dye molecules have three roles: They selectively absorb light from the color spectrum, they dissolve in water, and they bind to natural fabrics, especially wool. Some

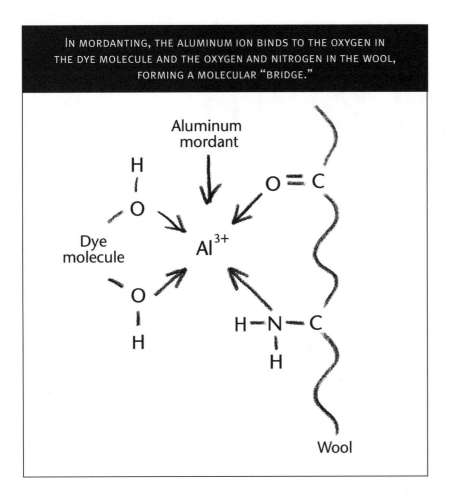

IN MORDANTING, THE ALUMINUM ION BINDS TO THE OXYGEN IN THE DYE MOLECULE AND THE OXYGEN AND NITROGEN IN THE WOOL, FORMING A MOLECULAR "BRIDGE."

dyes require mordants to assist them in binding.

The atoms in dye molecules are bonded together by electrons. In some of these bonds, the electrons move back and forth at a rate corresponding to the frequency of visible light. The exact rate determines the frequency of light that will be absorbed. The frequencies that are not absorbed determine the color that you see.

In an acidic solution, such as the dye bath you used, there is a diminished supply of electrons. The triply positive aluminum ion in the mordant attracts the negatively charged electrons of the oxygen atoms in the dye molecule and the negatively charged electrons of the oxygen and the nitrogen in the wool, linking them together. Chemists call the resulting links *ligands*.

The aluminum gathers and coordinates with four elements: three oxygens and one nitrogen. By forming ligands with atoms in both the dye and the wool, the aluminum mordant acts as a bridge between them.

This activity was developed by Kim Shuck.

Making Connections

- Does color play a part in the way people dress where you live? Do young and old people wear the same colors? Do boys and girls wear the same colors?

- Are there particular colors you like or hate to wear? Are there colors you wouldn't wear because of other people's reactions to them?

- Are any of the "rules" about colors of clothing written down? How do people know what the rules are?

- Have you ever dyed anything before? What did you make? How did you make it?

Recommended Resources

Dean, Jennie. *Wild Color.* New York: Watson-Guptill Publications, 1999.

Epp, Dianne N. *The Chemistry of Natural Dyes, Vol. 2* (Palette of Color Monograph Series). Middletown, OH: Terrific Science Press, 1995.

Liles, J. N. *Art and Craft of Natural Dyeing: Traditional Recipes for Modern Use.* Knoxville: University of Tennessee Press, 1990.

10

TEA AND TEMPERATURE

Chinese Traditions

TEA AND TEMPERATURE
Chinese Traditions

The temperatures of the food and drink we enjoy are important to us. In China, where people like their tea hot, they keep it from getting cold by using cups with lids on them.

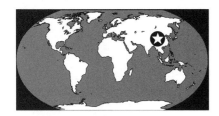

The First Cup of Tea

Both tea leaves and porcelain—the material that most fine teacups are made of—were introduced by the people of ancient China. It is said that tea was first made there about 5,000 years ago—by accident.

According to legend, the Chinese Emperor Shen Nong was concerned about hygiene and health in his country. He declared that all water must be boiled before it could be drunk. One day, while visiting a distant part of the country, the emperor stopped for a drink. His servants built a fire and began to boil some water. By chance, the fire was near a tea bush. A wind blew some dried tea leaves into the pot, creating a brown liquid with an appealing smell. The emperor was a man of science and experimentation, so he took a sip—and declared the new beverage delicious.

This story is part of an oral tradition, more myth than history, but probably parallels (to some degree) how the first cup of tea was really created.

THE CUP ON THE LEFT HAS A LID THAT CAN BE SHIFTED JUST ENOUGH FOR THE TEA DRINKER TO TAKE A SIP WITHOUT UNCOVERING THE WHOLE CUP.

The first recorded use of tea was in southern China in 50 B.C. After that, the popularity of the new drink spread rapidly through Chinese society. Tea was used as a medicine and a spiritual drink. In the eighth century A.D., a T'ang Dynasty scholar named Lu Yu wrote a book called the *Ch'a Ching,* or *The Classic of Tea,* in which he described the benefits of tea. Lu Yu's book tells everything about tea, from growing the plant to the proper cups for drinking it. The book brought Lu Yu fame,

and in his later years he was patronized by the emperor, who received tributes of tea from his subjects every year.

Teapots and Teacups

Tea was brought west by explorers who visited China, first from the Middle East and, hundreds of years later, from Europe.

By the early 1700s, tea was prized in England. Each year, clipper ships loaded with tea from China raced to reach England,

EIGHTH-CENTURY TEA MASTER LU YU.

THE CHINESE CHARACTER FOR "TEA."

Today, the Chinese use porcelain at home for their own daily meals and tea drinking, and export beautifully decorated porcelain all over the world.

The use of both tea and porcelain spread from China around the world hundreds of years ago. As tea traveled, new customs and traditions developed surrounding the ways it was served.

In most places outside of China, loose tea is brewed in lidded teapots and then poured into open cups or mugs. Rather than using covered pots, the Chinese put tea leaves into individual covered cups. Water is boiled, added to the cup, and then the lid is put immediately on. Since the brewing takes place right in the teacups, the host can adjust the strength of the tea to each guest's taste.

For tea to be brewed properly in the Chinese tradition, all the dry leaves must be wetted so they will sink into the water. The lids on Chinese teacups keep the rapidly evaporating water from escaping. The water that evaporates into steam stays in the space between the tea and the lid. That steam helps wet the floating leaves, so the drinker can enjoy a good cup of tea.

where the first tea of the season would bring an extremely high price. Many of the tea ships carried no other cargo because the captains would not allow anything that might spoil the tea with dampness or odor. Even the ballast (material carried as weight to steady the ship) had to be carefully considered. Many captains used boxes of porcelain (plates, cups, and other fine tableware) as ballast because the porcelain could not harm the valuable cargo of tea.

Porcelain was also developed in China, in the third century A.D. It came from large natural deposits of *petuntse* (also called *china stone*, a kind of feldspar) and *kaolin* (also called *white china clay*). The Chinese traded porcelain to markets in Asia during the next few centuries. When porcelain reached Europe in 851, people were astonished by its delicate quality and translucence. The cups and plates used in Europe at that time were made of thick, coarse earthenware.

WHAT'S IT ALL ABOUT?

Both activities in this investigation use teacups—with and without lids—to observe and investigate how heat is lost from hot water. As you experiment, you'll also find out about the centuries-old tradition of tea in China. You'll discover:

- How well lids work to keep tea hot
- What happens to make a cup of hot tea lose its heat
- What conduction, convection, and radiation are

Activity 1: Conduction, Convection, and Radiation

In this activity, you'll learn about the three different ways heat is transferred: conduction, convection, and radiation.

Activity 2: Heat Loss and Heat Capacity

You'll measure heat loss and think about a basic scientific question: What is heat?

The lids of Chinese teacups have little knobs in the center. They are lifted slightly each time a drinker takes a sip of tea. It's the host's job to keep the cups filled. The level of the tea in a guest's cup is rarely allowed to drop below an inch from the top, even during several hours of conversation and tea sipping.

There is another reason for using lidded teacups: The Chinese, like many people, enjoy drinking their tea slowly and want to keep it warm for as long as possible. Central heating is rare in China, and in winter the temperature in many houses is near freezing. The lids help keep the tea hot in the cups.

Activity 1: Conduction, Convection, and Radiation

In this activity, you'll observe and measure the temperature in different parts of cups filled with boiling water. You'll also learn about the three ways heat is transferred: by conduction, convection, and radiation.

Conduction

To begin this activity, fill a teacup with boiling water to about half an inch below the brim. Then quickly cover the cup with a lid.

After a few seconds, use your fingers to carefully feel different parts of the cup. Touch the sides, bottom, handle, and lid. Be careful not to spill any of the water on yourself! ⚠

Where does the cup feel the hottest? Where does it feel the coolest? 🖊

You'll probably find that the coolest places are on the handle and lid of the cup, and the hottest places are on the sides and bottom of the cup. Why do you think the temperature is distributed this way? 🖊

The hottest places on the cup are where the most heat is being lost. Cooler places are losing less heat. At the top of the cup, the water does not touch the lid, so there's a pocket of air between the water and the lid. The heat must pass through that pocket of air and then through the ceramic lid before it can get to your fingers. At the sides, though, the heat in the water only has to pass through the cup to get out to where you can feel it.

It is scientifically correct to say "Heat is trying to escape from the water." Heat is flowing from the water to the teacup, and then to the air and to the table on which the cup sits. When you touch the teacup, heat flows to your fingers.

The transfer of heat through a material is called *conduction*. Conduction is one of the three basic ways heat is transferred. When the lid is on the cup, most of the heat leaving the water is escaping through conduction. Air is less efficient than the ceramic cup at transferring heat, so less heat is

A TRAFFIC CONTROLLER IN BEIJING ENJOYS HIS DAILY TEA, KEPT WARM IN ITS LIDDED CUPS.

ACTIVITY 1: CONDUCTION, CONVECTION, AND RADIATION

Is there anything special I should know?

- This activity is recommended for ages 12 and up
- For safety reasons, it's a good idea to work on this activity in groups of two—that way, it's easier to avoid any spills of the hot water you'll be handling
- If children are doing this activity, an adult should be supervising their work: Boiling water can cause serious burns! ⚠
- If possible, participants should wear protective goggles.

How much time will I need?

- This activity takes about 1 to 2 hours

What materials will I need?

- Ceramic cups or mugs with handles
- A ceramic lid for each cup (if you can't find lidded teacups, you can use glass or ceramic saucers as covers)
- Thermometers that can measure temperatures up to 100°C (212°F)
- Boiling water
- Masking tape
- A ruler marked in centimeters
- A few drops of any type of cooking oil
- Scissors
- Graph paper

transferred to the lid than to the outside walls of the cup.

Heat always flows from a hot place to a cooler place. This fact is the basis of the Second Law of Thermodynamics (the part of physics that deals with heat energy). It states: *Heat energy flows spontaneously from a substance at a high temperature to a substance at a low temperature, and does not flow spontaneously in the reverse direction.*

You've probably experienced this yourself. Leave your hot chocolate sitting out and it doesn't get hotter, it gets cooler. Things don't spontaneously heat up.

What you have observed so far

gives you a *qualitative* measure of temperature; that is, you know the *quality* of being hot or cold. You can make comparisons between the different temperatures to find out where things are hotter and where things are cooler. Suppose you want to know, though, exactly how hot a point on the cup is? Without

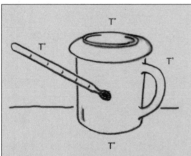

You can quantify the temperatures in different places on a hot-water-filled cup more accurately with a thermometer. Try holding the thermometer's mercury bulb against the side of the cup. (Wet the bulb with a drop of oil to improve contact with the cup.) Wait until the mercury stops rising, then write down the temperature. Try this at four points on the cup and write down each of your measurements.

The transfer of heat from the cup's surface to the thermometer's glass bulb will not be very efficient. It takes patience and steady hands to hold the thermometer still long enough for the level of the mercury to stabilize.

People who have tried this activity have found that there is about a six-degree difference in temperature between the side of the cup and the lid, and no difference between the side and the bottom.

using a thermometer, how could you "measure" the temperature at different points to get a *quantitative* measure?

One way to measure the temperature is to use your sense of touch. Earlier, you found that when you touched the cup, the lid was slightly cooler than the sides or the bottom; the handle was probably the coolest place of all. One concrete way to check this observation is to press your fingers on the sides of the cup and count how many seconds pass until you are uncomfortable and have to move your fingers. Write down that number. When your fingers cool down, try the same thing on the lid and record that number, too. The lower the number, the hotter the cup.

Remember, the point of this exercise is to compare the temperature in different places—not to see how long you can stand to touch a hot cup. Don't burn yourself! ⚠

Convection

If the water has cooled off, refill your cup with fresh hot water and put the lid on top. Then wait a minute or two to start this activity so the steam doesn't burn your hand. ⚠

When you're ready, carefully remove the lid and check on the temperature by holding your fingers near the cup, but not touching it. Can you feel the heat around, below, and above the cup? Where do you feel the most heat?

The hottest place, by far, will be right above the cup. If it feels as if the heat is rushing onto your hands, you're right. It may surprise you to know that gravity is responsible for this rush of heat.

Air expands as it gets warmer. Since its molecules move farther apart, warm air becomes less dense than cool air. Gravity pulls down

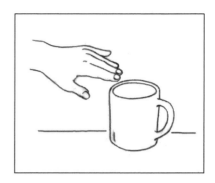

harder on the denser, cooler air, and the warmer air rises. This principle is called *convection*.

In the teacup, the water heats the air just above it, and that air rises. The rising hot air is replaced by cooler air coming from the sides, as shown in the illustration below. This "new" air quickly becomes heated, and then it rises. Whenever you see water vapor rising, you're watching convection at work.

You can tell with your fingers where the air is warmer and cooler around the uncovered cup. You may want to try to quantify the effects of convection by taking the temperature of the air around the teacup at varying distances. How would you do that?

One way is to attach a thermometer to a "ruler" made of masking tape, as shown on page 111.

First, stick a strip of masking tape 14 centimeters long just above the bulb of the thermometer. Double the tape back on itself, sticky side to sticky side. You'll end up with a strip of tape about 6 centimeters long attached to the thermometer. Use a ruler to mark off the tape in centimeters.

Now hold the tape so that it's perpendicular to the side of the cup and measure the air temperature 6 centimeters from the cup. When the mercury stops rising, write down the temperature. Then take the temperature 6 centimeters above the cup. Write that down, then measure the temperature 6 centimeters below the cup and record that, too.

When you have three measurements at the 6-centimeter distance, cut the tape off at the 3-centimeter mark and measure the air temperature at the same three locations, 3 centimeters from the cup. Then cut the tape off at the 1-centimeter mark and make three more measurements. Record each of your measurements as you make them.

Now take a look at the temperatures you have recorded. What do they tell you about the way the heat flows away from the water? When the lid is off, is more heat convected up out of the cup or is more heat conducted through the sides? ●

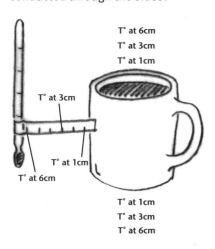

T° at 6cm
T° at 3cm
T° at 1cm

T° at 3cm

T° at 1cm

T° at 6cm

T° at 1cm
T° at 3cm
T° at 6cm

The air above each teacup will probably be the hottest, due to con-

MING DYNASTY (FOURTEENTH CENTURY) OFFICIALS ENJOYING ONE OF THEIR MANY DAILY TEA BREAKS, FOLLOWING THE OLD SAYING, "TEA SHOULD BE DRUNK OFTEN BUT IN SMALL QUANTITIES."

vection. The air at the sides should be cooler, and the air at the bottom should be almost the same as at the sides.

Radiation

The third method of heat transfer is *radiation*. If you put your fingers near the sides of the cup, you feel radiated heat. You also feel radiated heat when you sit in the sun or next to a heater. The cup, like the sun and the heater, is radiating heat in the form of *infrared radiation*.

All hot things transfer heat by radiation. Infrared radiation from the sun is part of the electromagnetic spectrum, as are radio waves, visible light, ultraviolet waves, and X rays. These waves differ only in their frequency.

Heat from the teacup is conducted

by the ceramic and the air and convected by the moving air. But infrared radiation doesn't require any material to transfer heat. The sun does a fine job of heating the earth through empty space. Unfortunately, there isn't any way to demonstrate this using the teacup. Convected heat rising from the top of the cup feels just as hot on your hand as radiated heat does from the side of the cup. But in one case, air is needed for heat transfer, and in the other, it's not.

Activity 2: Heat Loss and Heat Capacity

What Is Heat?

Now that you have experienced the three ways that heat is transferred, what do you think heat is?

Try making a map of the temperatures that surround a cup filled with hot water. You've already recorded the temperatures 1, 3, and 6 centimeters from the sides and bottom of the cup. Can you find the position (or positions) above the cup where the temperature is the same as it is 6 centimeters from the sides and bottom? ☝

Now draw a map connecting these points around the cup. When you connect the points of equal temperature, you are drawing *isotherms*, lines of equal temperature. ☝

Where else have you seen isotherms? Take a look at weather maps in the newspaper or watch the weather report on your local TV news. You'll see isotherms around areas where the temperature is the same. Isotherms can be used to "map" heat flow. ☝

What is this stuff that was once in the water and has escaped into the room? ☝

Scientists have only recently come to understand what heat is. Throughout history many people have thought that heat was a fluid—a material thing.

We now understand that heat is not a substance at all. It is energy. Heat is the level of agitation and internal excitation of the molecules that make up a substance. Water is hot when the water molecules bounce against one another rapidly. Water gets colder when those same molecules slow down. Temperature is the measure of this molecular agitation, as detected with a thermometer.

Heat Loss

Now that you have an idea of where the heat from the water is going, you can do an experiment to see how fast the heat is lost.

Carefully fill two cups to the brim with boiling water and fill the other two cups only one-quarter full of boiling water. Put a lid on one of the full cups and on one of the quarter-full cups, as shown on the bottom of page 113. ☝

To see how fast heat is lost from the cups, take the temperature of the water in each of the four cups every 5 minutes for 45 minutes. (For the lidded cups, just slide the lid to the side enough to accommodate the thermometer.) Be sure to write down the measurement for each cup as you take it. You may want to devise a table that will help you keep track of the measurements as you work.

Before you start, think about this: Based on what you have observed so far about heat flow, which cup do you think will lose heat most quickly? Which will lose heat most slowly? Will the heat loss be steady, or will it speed up or slow down after a while? ☝

When you have all your data, look carefully at the results. Make a graph of the temperature change in each cup. Use the *x*-axis for time and the *y*-axis for temperature. What do your graphs show you? (See sample on page 114.)

You'll discover that the temperatures drop steadily, but at a different rate for each cup. Subtract the most recent temperature reading for a cup from the first reading for that cup to get the temperature drop during that time period. ☝

The full cup with the lid will have the least temperature drop. This is what the Chinese have known for centuries—using a full cup with a lid keeps the tea warm longer.

ACTIVITY 2: HEAT LOSS AND HEAT CAPACITY

Is there anything special I should know?

- This activity is recommended for ages 12 and up
- You can work alone or with friends
- As with the other activity in this chapter, be very careful with boiling water and wear goggles if they are available. ⚠

How much time will I need?

- This activity takes 1 1/2 to 2 hours if you have 4 thermometers

What materials will I need?

- 4 identical (or very similar) cups
- 2 identical (or very similar) lids for the cups
- 1 to 4 thermometers (if you use different thermometers, check them in one cup of hot water to be sure they are calibrated the same)
- Boiling water
- Graph paper
- Semilog graph paper (optional)

After about 40 minutes, the difference in temperatures between the two quarter-full teacups becomes very small. This is probably because both partially filled cups have a large air pocket above the liquid. Even in the cup without a lid, the pocket of air—often called an air cell—helps keep the heat in the liquid. (Remember that air is less efficient at conducting heat than the ceramic of the cup is.) The heat that escapes is lost mainly by conduction through the walls of the cup.

Heat Capacity

Different materials absorb heat in different ways, so their temperatures rise at different rates. Suppose you had one quart of oil and one quart of water, both at room temperature, and you wanted to heat them both to 100°C. You'd have to leave the water on the fire about twice as long as the oil to get the same rise in temperature. The *heat capacity* of water is twice as high as the heat capacity of oil.

Can the teacup itself hold heat energy, just as the water can? Try this experiment:

Fill a cup with hot water. Wait a few minutes and then pour out the water. Rinse the outside of the cup in cold water for a few seconds, then grab the cup and hold on. ◗

You'll notice that the cup will feel cool at first, and then the heat from the inside of the cup will spread to the surface again. This is because the cup itself has heat capacity, although it has less than the water.

Many cultures have incorporated the practical uses of heat loss into their rituals. While people who live in the coldest parts of China drink their tea in lidded ceramic cups to keep the heat in, people who live in the hot, humid climate of tropical India drink out of metal cups without lids because they want their tea to cool off as fast as possible.

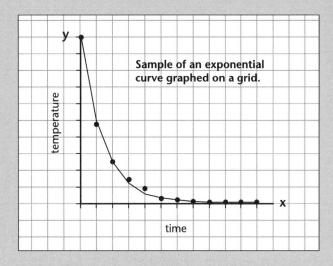

Sample of an exponential curve graphed on a grid.

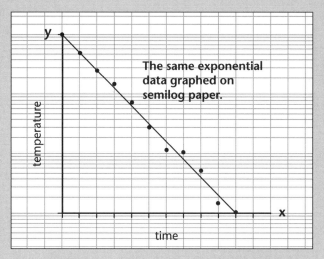

The same exponential data graphed on semilog paper.

The Mathematics of Cooling

What do these graphs represent? The English physicist Isaac Newton pondered this question more than three hundred years ago. In 1701, he experimented with heat loss, using a red-hot piece of iron instead of teacups. Curves similar to the ones in your heat-loss experiment appeared on Newton's graph. He found that they were *exponential decay curves*. This means that as time passed in an arithmetic progression, the temperatures decreased in a geometric progression.

An arithmetic progression is a sequence of numbers in which the same number is added each time to get the next number. For example, 0, 2, 4, 6, 8, . . . (0 + 2 + 2 + 2 + 2 . . .).

A geometric progression is a sequence of numbers in which each number is multiplied by a constant to get the next number. For example, 1, 2, 4, 8, 16, . . . (1 x 2 x 2 x 2 x 2 . . .).

The curves on this page show that the time is passing arithmetically as the temperatures are moving geometrically. The loss of heat is exponential. This is called Newton's Law of Cooling.

Newton wrote:

If the times of cooling are taken equal, the heats will be in geometri-cal progression. . . . By assuming that the excess of the heat of the iron and of the hardening bodies above the heat of the atmosphere, found by the thermometer, were in geometrical progression when the times were in arithmetical progression, all the heats were determined.

I placed the [very hot block of] iron not in quiet air but in a uniformly blowing wind, so that the air warmed by the iron would be continually taken away by the wind, and cold air would come in its place with a uniform motion. For thus equal parts of the air are warmed in equal times and carry away a heat proportional to the heat of the iron.

How can you be sure that the curve is an exponential? There are two ways. One way is to take the logarithm of each temperature and graph the logarithms against time. (A logarithm is the inverse of an exponential. You can easily find each temperature's logarithm by using a scientific calculator with a logarithm function, or you can do what Newton did and look up the logarithms in a table.)

The other way to find out if a curve is an exponential is to use semilog graph paper. Semilog graph paper takes an exponential curve and transforms it into a straight line. In one direction, the

lines on the grid are not evenly spaced—they are pushed together at one end and pulled apart at the other—following an exponential function. In the other direction, the grid lines are evenly spaced, as on regular graph paper.

Because the temperatures drop geometrically, plot the temperatures on the unevenly spaced lines of the semilog paper. The times, which increase arithmetically (every five minutes) would be plotted on the evenly spaced lines. If the curve is indeed an exponential, you'll get approximately straight lines.

If you plot the data from your teacup experiment, you'll find that the line for a lidded cup is less steep than it is for an open cup. You've discovered that the rate of temperature drop, and therefore of heat loss, is slower for the lidded cup, which is what Chinese tea drinkers take advantage of.

To get a drastically different illustration of heat loss, carefully fill an aluminum soda can with boiling water and repeat the same experiment. The line on the semilog graph paper will be much steeper than the line for the teacup, because heat conduction through the aluminum walls of the can is much greater than through the ceramic walls of the teacup.

This activity was developed by Curt Gabrielson.

Making Connections

- Does your family regularly drink tea, or is there another drink—hot or cold—that's traditional to your country or culture?

- People in the cool parts of China like to keep their tea hot; people in the hot climates of India like to let their tea cool. Do you eat or drink special things when the temperature is hot or cold? How do you keep it at the temperature you like it?

- How is your body like a cup of hot tea? How do you keep yourself warm on cold days or cool on hot days?

Recommended Resources

Newton, Isaac. "A Scale of the Degrees of Heat" (1701) in William F. Magie, *Sourcebook in Physics.* Cambridge: Harvard University Press, 1963.

Wood, Robert W. *Physics for Kids— 49 Easy Experiments with Heat.* New York: Tab Books, 1990.

Yu, Lu. *The Classic of Tea: Origins and Rituals.* Translation of the *Ch'a Ching* with an introduction by Francis Ross Carpenter. Boston: Little, Brown and Company, 1974.

SUBSISTENCE AND SURVIVAL

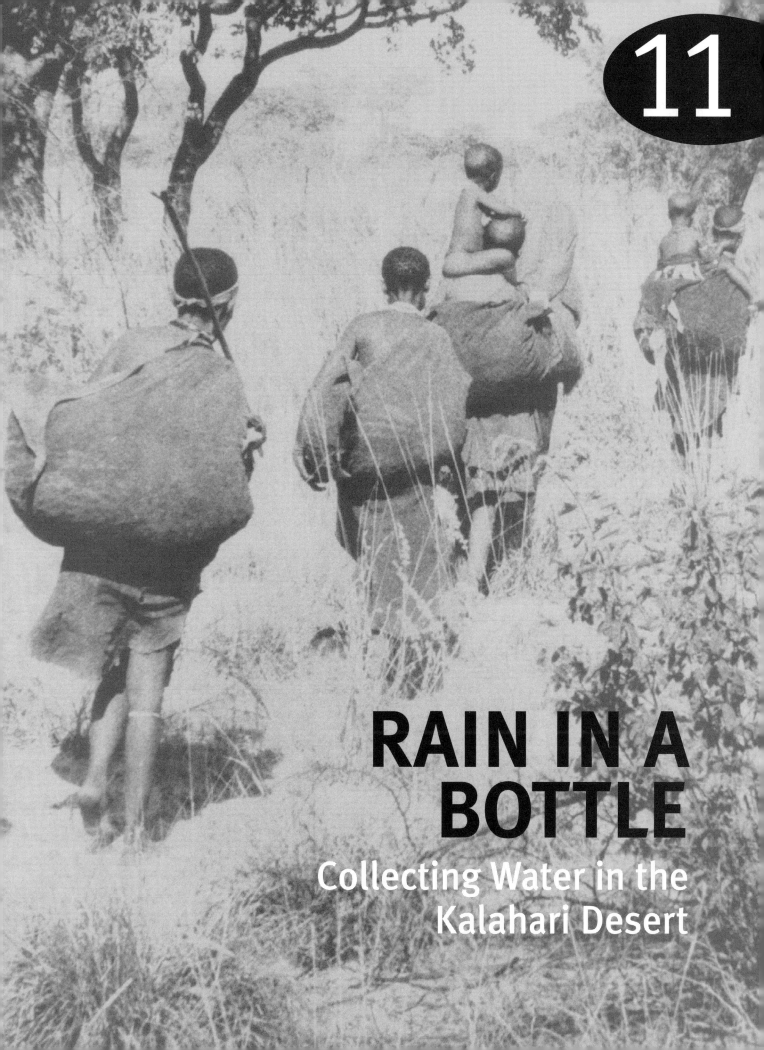

11

RAIN IN A BOTTLE

Collecting Water in the Kalahari Desert

RAIN IN A BOTTLE

Collecting Water in the Kalahari Desert

In some parts of the world, people worry about having enough water to fill up their swimming pools. In the desert, finding water may be a matter of life or death.

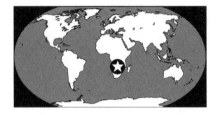

A Little About the Basarwa

The Basarwa are an ancient tribe of people originally called the *San*, or "gatherers," and sometimes called Bushmen or !Kung. (The ! in !Kung stands for a popping sound made with the tongue, a regular sound in the Basarwa language.)

The Basarwa live in southern Africa in one of the world's harshest deserts—the *Kgalagadi*. (The name is pronounced "Ka-la-ha-di" in the Basarwa's Setswana language; in English, this desert is known as the Kalahari.) Created by the erosion of ancient volcanoes, the sands of the Kalahari lie between two mountain ranges along the eastern and western borders of Botswana.

Flowing into this giant sun-baked sandbox is the Okavango, a river that once fed a huge inland sea. The river's path has changed over the years. Today it simply disappears along Botswana's northern border into the Okavango Delta. This lush place is one of the last wild areas left in Africa, where big mammals and hundreds of species of birds share the limited food and water.

A BASARWA HUNTER DRINKS FROM A WATER HOLE USING A CUP MADE FROM A CUCUMBER.

A BASARWA WOMAN HOLDS TWO OSTRICH EGGS USED AS WATER CANTEENS.

The existence of all life in the Kalahari Desert is inseparably linked to water, and who or what gets it. In the African summer—mainly in January, February, and March—rain falls in fierce thunderstorms that may drench only one small area. During the rainy season, part of the desert called the Makgadikgadi Salt Pan (a pool that leaves salty deposits behind when it dries up) becomes a shallow lake, almost a meter deep. After the rains, the hard and impermeable lake bed—which is made up of calcium carbonate, or "hard pan"—helps hold water in pools scattered all over the desert. This allows animals and people to move freely across the vast lake bed without being dependent on a single location as a source of water.

For thousands of years, the nomadic Basarwa have moved across the Kalahari from one remembered water site to another before the resources in any one area were used up or soiled. Their entire world has always revolved around the cycles of nature. By capturing the water they need in varied and creative ways, they can survive through decades with little or no rain.

To get some of their water, the Basarwa dig up tubers—thick, fleshy roots that some plants grow under the ground. They don't eat the tubers; instead, they shred them and squeeze out the liquid to drink. The Basarwa also gather water from hollow trees and from water holes, where underground water comes to

the surface. Some water holes are good only in the rainy season, and some always have water.

A Basarwa Method of Collecting Water

When water is scarce in their dry desert home, the Basarwa people of the Kalahari collect water by trapping moist air and condensing the water out of it. First they dig a small pit in the sand and lay grass in it. They place half of the shell of an ostrich's egg in the center of the pit, open side up, like a cup. Then they stretch a transparent animal membrane (usually a bladder) over the pit, holding the edges down with stones. They put a smaller stone in the middle of the membrane to

121

weigh it down so that it sags into a cone shape. The point of the cone is aimed down into the eggshell.

During the day, the sunlight's energy passes through the transparent membrane and heats up the grass and surrounding sand. The small amount of water in the grass and sand starts to evaporate into the air under the membrane. The pit becomes a small container of humid air.

At night, the cold desert air cools down the membrane. As the humid air under the membrane cools, water condenses, running down the bottom of the membrane and dripping into the eggshell. This technique can produce up to a quarter-cup of water in one night. By digging several holes in one area, the Basarwa can collect enough water to survive in the Kalahari.

The Basarwa also use ostrich eggshells to carry water that has been collected. They drill a hole in the shell, fill the shell with water, and plug the hole with a stopper of grass. Filled with water, the shells weigh about three pounds.

Building Your Still

Follow these steps to make a working model of a Basarwa still. You might want to make your still in the morning, since you may have to leave it overnight.

1. Cut the 2-liter bottle in two, at about two-thirds of its height.

2. To create brackish water, pour about half a cup of tap water into the bottom part of the bottle. Add about 1 teaspoon of salt and two

WHAT'S IT ALL ABOUT?

In this activity, you'll re-create one of the methods people all over the world have used to collect and purify water. Using easily available materials, you'll start with brackish (salty), dirty-looking water. Then you'll make a model of a still—a simple device for purifying water—and observe the processes of evaporation and condensation as your still collects clean water. As you work, you'll discover:

- How the Basarwa people of the Kalahari find water in the desert
- How evaporation and condensation can be used to collect and purify water
- How changes in Africa are affecting the traditions of the Basarwa today

Is there anything special I should know?

- This activity is recommended for ages 10 and up
- You can work alone or with others
- Before you begin, you should be familiar with the parts of the water cycle—rainfall, evaporation, and condensation—and you should be able to imagine what atoms and molecules are

How much time will I need?

There are three parts to this activity:

- Constructing a Basarwa still should be done in the morning; it takes about an hour

- The water purification portion of the activity may take several hours, or overnight
- Testing the purified water takes about half an hour

What materials will I need?

- 2-liter transparent plastic soda bottle with its cap
- Plastic cup that will fit inside the bottom of the bottle
- Scissors
- Food coloring set (4 colors)
- Water
- Salt
- A few marbles
- Lamps or heaters, in addition to heat from sunlight (optional)
- Ice cubes (optional)
- Plastic turkey baster (optional)
- Aluminum foil
- Stove or other heating element

BASARWA WOMEN DIG UP TUBERS, THEN SQUEEZE OUT THE WATER INSIDE.

membrane

stones

grasses

ostrich egg half

sand

A CUTOUT VIEW OF A BASARWA STILL

drops each of red, yellow, blue, and green food coloring. The food coloring makes the water darker, which helps it absorb energy as heat. The greenish-brown color also makes the water look undrinkable.

Stir or swish the solution around to dissolve the colors and the salt.

3. Place a small plastic cup in the bottom part of the bottle. Wash and rinse two or three marbles and put them in the bottom of the cup to weigh it down so it does not bob around in the brackish water.

4. Make sure that the plastic cap is screwed tightly onto the top part of the bottle. Turn the top part upside down and set it inside the bottom part, as shown below.

Make sure that the bottle neck points down toward the collection cup but does not touch the sides of the cup. You want the drops of condensed water to run down the neck of the bottle and drip into the cup. You may need to move the top of the bottle around until the neck is lined up above the middle of the cup without touching the sides.

5. Once the top of the bottle is arranged inside the bottom, push it down gently, just a little, to make sure that it fits firmly and forms a tight seal. Otherwise the evaporating water will escape.

Making Your Still Work

Take a close look at your still. How are the parts of your still like the parts of the stills the Basarwa use?

You may have noticed that the bottom part of the bottle is like the pit in the sand. The cup inside it is like the ostrich eggshell. The brackish water is the moisture in the grasses inside the pit. The upside-down top is like the transparent membrane stretched over the hole.

Watch what happens as your still purifies the brackish water. Over time, water will evaporate from the colored liquid, condense on the underside of the top of the bottle, and drip from the cap into the collection cup. There's one problem: This process may be very slow.

In the desert, the climate pattern is daytime heat and nighttime cold. This means that evaporation takes place during the day, and con-densation takes place at night. Will this happen where you've set up your still? Depending on where you're located, and what time of year it is, there may be little sun during the day to heat up the brackish water or little cold air at night to cool the top.

Can you think of ways to speed up evaporation and condensation so that you can collect water in your still more quickly?

One way is to heat up the brackish water to increase evaporation. That's one reason you might decide to put your still outside, in direct sunlight. You can also put it near a heater or heat lamp. Placing the whole still on a piece of black paper may help increase the heat it absorbs.

To speed up condensation, you might put three or four ice cubes in the inverted top of the bottle to cool off the plastic walls. Screw the cap on tightly to make sure that the water dripping into the collection cup comes from condensation, not from melting ice leaking through the cap.

Try to observe the still every half hour. Take notes on what you see. Can you see any evidence that the

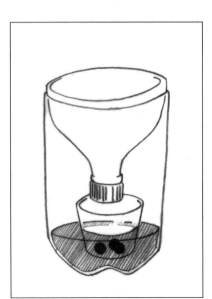

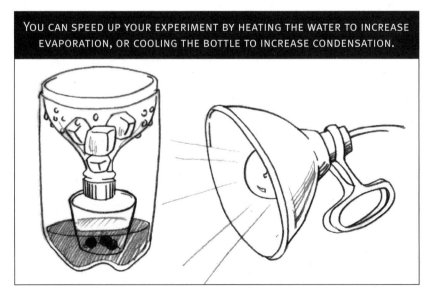

YOU CAN SPEED UP YOUR EXPERIMENT BY HEATING THE WATER TO INCREASE EVAPORATION, OR COOLING THE BOTTLE TO INCREASE CONDENSATION.

still is working? If you're using a heater and ice cubes, you can keep replacing the ice as it melts if you want to speed up the process. Just suck out the extra water with a turkey baster.

Testing the Water

After a few hours, see how much water has collected in the cup. If only a few drops have collected, leave the still to work overnight. When there is enough water to taste, carefully remove the top part of the bottle and dump any melted ice into the sink. (Make sure none of the clear water from the melted ice accidentally spills into the collection cup!) Then remove the collection cup from the bottom of the bottle.

What color is the water in the collection cup? What does it taste like? Can you taste any salt?

There is another way to test the purity of the water you collected. Put a few drops of the water from your collection cup on a piece of aluminum foil. Ask an adult to place the foil on a stovetop burner and let the water evaporate. This should take only a few seconds.

Now put a few drops of the brackish water from the bottom of your still on another piece of foil and evaporate it in the same way. What do you see?

After it evaporated, the brackish water probably left a small deposit of salt on the foil. But the drop of pure water from the still most likely left no visible mark.

CHANGES IN THE KALAHARI DESERT

Until recently, the Basarwa's way of life was protected by two things: the tsetse fly in the Okavango Delta, which carries sleeping sickness, and the Kalahari Desert itself. The disease and the lack of water once limited the number of cattle that could live in the region. Beginning in the 1980s, however, experts with specialized tools and technology brought the tsetse fly under better control and drilled deep wells to bring clean, fresh water from deep underground to the surface of the desert.

As these problems were solved, many more people began moving into the area. Some came from Europe and other distant cultures; others came from nearby, including Bantu cattle herders from Botswana. For the Bantu, who are the largest population in the area, cattle represent wealth; for the European markets, cattle represent food. This combination has created a tremendous demand for grazing land, destroying the sustainable ecology in which the Basarwa lived.

With less open land available to them, the Basarwa's nomadic lifestyle is being destroyed, and they are being forced to fit in with the majority culture. Few Basarwa have survived untouched by these changes. Slowly, the traditional knowledge and culture of the Basarwa are disappearing into the dry sands.

Exploring the Chemistry of Evaporation

To make brackish water, you combined tap water, salt, and food coloring. Why do you think the water ended up in the collecting cup, but the salt and the coloring did not?

You probably know that when water evaporates from the ocean, it falls back to earth as rain. The rain is fresh water, not salt water. The same thing happens in the still.

Water consists of small, light molecules made up of two hydrogen atoms and one oxygen atom (H_2O). The food coloring is made up of bigger, heavier molecules. When the brackish liquid warms up, all the molecules move faster. But the heavier molecules move more slowly than the lighter ones.

In evaporation, molecules escape from the surface of the liquid into the air. In your still, the water molecules reach the surface more often and faster than the heavier molecules of the food coloring.

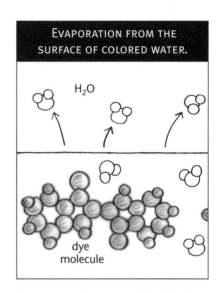

EVAPORATION FROM THE SURFACE OF COLORED WATER.

H_2O

dye molecule

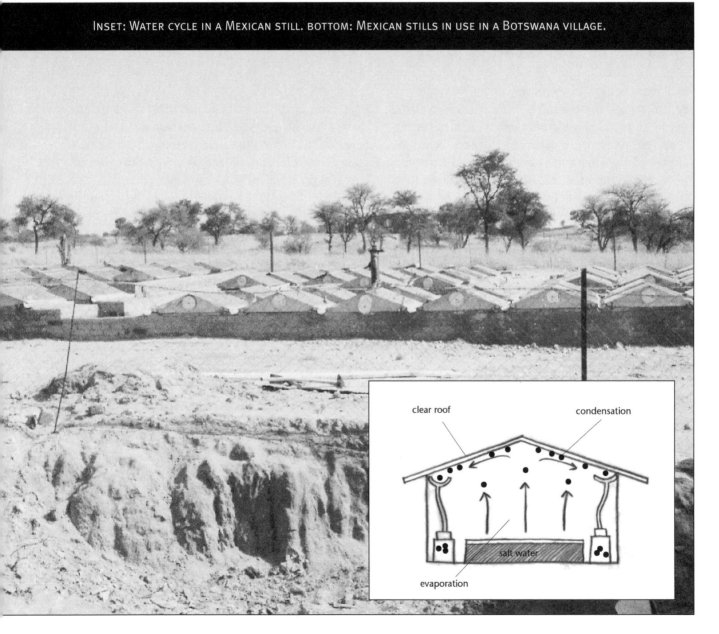

INSET: WATER CYCLE IN A MEXICAN STILL. BOTTOM: MEXICAN STILLS IN USE IN A BOTSWANA VILLAGE.

clear roof

condensation

salt water

evaporation

The result is that the contaminants—dirt, salt, and coloring—are left behind, while the water molecules escape into the air. When the water molecules cool and condense, the water that drips into the cup is fresh and clean.

Real-Life Applications

The water-purifying technique demonstrated by your still is also used in the dry desert regions of Mexico. The "Mexican still," which can collect much more water than a traditional Basarwa still, is now in use in other parts of the world, including the Kalahari.

A Mexican still is shaped like a tent. Inside, there's a long tray filled with the brackish local water. The roof is transparent. Heat from the sun evaporates the brackish water, and the cold night air makes the evaporated water condense on the ceiling of the tent. The water rolls down the tent walls and drips into gutters at the sides of the structure. From the gutters, the water drains down into one of several collection tanks. This process produces fresh drinking water. After the brackish water evaporates, the salt that is left behind is sold in local markets. A two-meter-square transparent roof (an area of about 36 square feet) yields about one liter of fresh water a day.

This activity was developed by Chris Andersen.

Making Connections

- What would it be like to live in a place where water isn't easily available? How do you think your life might be different? How might it be the same?

- The Basarwa live a nomadic lifestyle, moving from place to place to find food and fresh water. Does your family move a lot, or have you lived in the same place all your life? If you could pick a new place to live, what kind of place would it be?

- Imagine being lost on an island in the middle of the ocean with no fresh water to drink. Based on what you now know about collecting and purifying water, could you think of ways to use what might be on the island to make or find clean water?

Recommended Resources

Shostak, Marjorie. *Nisa: The Life and Words of a !Kung Woman*. Cambridge: Harvard University Press, 1983.

Smith, Andrew, Candy Malherbe, Matt Guenther, and Penny Berens. *The Bushmen of Southern Africa: A Foraging Society in Transition*. Athens, OH: Ohio University Press, 2000.

Van Der Post, Laurens. *The Lost World of the Kalahari*. New York: Harcourt Brace, 1977.

12

FROM ARROWS TO ROCKETS

Flight Stability

FROM ARROWS TO ROCKETS Flight Stability

Launching something through the air so that it reaches its target can be quite difficult. Today we take on this challenge in a variety of games where hitting a target is a valuable skill. But in the past, and in some parts of the world today, hitting a target could make the difference between eating and going hungry.

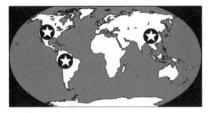

Staying on Target

In different cultures at different times, hunters have developed specialized tools so they could hit distant targets and get the food they needed. Aborigines in Australia developed the boomerang. It could sweep across the empty desert and, if it missed its target, come back in another wide sweep to the hands of the hunter. The boomerang saved the hunter from having to track down the weapon before it could be used again.

The boomerang would not be as useful in a forest, however, where it would be knocked down by trees. Native people who lived in the wooded areas of North and South America developed different kinds of projectiles, including darts and blowguns and bows and arrows. Narrow, straight-flying darts and arrows were especially good for passing between bushes and trees to reach their targets.

Darts that are used in the popular game today were developed in

A YOKUTS MAN ATTACHES AN ARROW-TIPPED FORESHAFT TO THE LONG BODY OF AN ARROW.

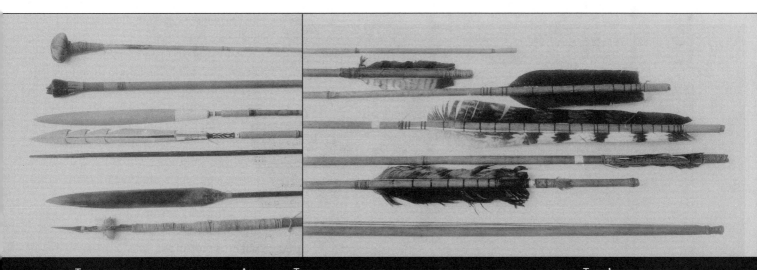

THESE ARROWS ARE FROM THE AMAZON. THE TOP TWO ARROW TIPS ARE BLUNT OR BALL-SHAPED. THEY'RE DESIGNED TO STUN BIRDS WITHOUT TEARING THEIR FLESH. THE NEXT FOUR ARROWHEADS, SOME WITH NOTCHES, ARE MADE FOR HUNTING SMALL GAME. THE ARROWHEAD ON THE BOTTOM IS FOR SPEARING FISH. IT HAS A DETACHABLE BARBED POINT THAT IS TIED TO A LONG LINE SO A FISHERMAN CAN PULL IT IN.

WHAT'S IT ALL ABOUT?

The four activities in this chapter can help you explore the way things fly. You can do all of them, or choose your favorites to discover:

- Why people put feathers on arrows
- How to calculate the altitude reached by a model rocket
- What physical forces affect how arrows and rockets fly
- How to "aim"—and hit a target—with a blown-up balloon

Activity 1: All About Arrows

Look at different kinds of arrows and compare their designs and uses.

Activity 2: Launching Rockets

Launch model rockets and watch them fly. If you want to know more, you can make a device called an "inclinometer" to measure how high the rockets go.

Activity 3: Flight Stability

Using a yardstick, you can learn about center of mass and center of pressure, two important characteristics of flying things.

Activity 4: Building Better Rockets

Design a balloon rocket that flies in a straight line and can hit a target.

the Middle Ages to help train English archers. They have fins like the feathers on an arrow. Europeans later developed rifles and bullets, then rockets. The Chinese made firecrackers that would shoot straight up in the air before they burst, so onlookers could watch in safety. Throughout history, people have used great technical ingenuity to make projectiles that would fly straight and true.

Native American Arrows

In Native American cultures that depended on hunting for food, spears and bows and arrows were carefully made for precise uses. Bows were made of strong, straight-grained wood, sometimes with deer sinew layered on. Simple, light bows were used for small game and the stronger, sinew-layered bows were used to hunt larger animals. Different kinds of spears and arrows were designed for different purposes, and

This arrow from the Amazon has only two half-feathers, each from the same side of a whole feather.

Arrows of the Chumash

At the back of the shaft, three half feathers are attached radially with sinew and asphaltum or pine pitch. Hawk tail feathers are preferred. They are split, and only the feathers from one side of the bird's tail are used, and always the same side of each feather. The feathers, after being split, are cut into six-inch lengths with a short stub of quill sticking out. Warm pitch is smeared on the arrow, and the feather carefully laid straight along the stem. The front end is tied with sinew first and then the back end. It is wrapped in these two places only. There is no spiraling of the three featherings. They are trimmed even by singeing with a live coal.

—from *Handbook of Yokuts Indians*, edited by Anne H. Gayton, 1929

Northern California Arrows

[The native people of the area] made arrows out of the straight shoots of the redberry bush. The arrows were 75 to 85 centimeters long and 7 to 9 millimeters thick. The diameter decreased a little toward one end. There was a small notch in the lower end of the arrow for the bowstring, and also three hawk or falcon feathers, split lengthwise, were attached at this end. The feathers, about 10 centimeters long, with trimmed edges, were tied at their ends to the shaft with thin strips of tendon.

—Karl von Loeffelholz, descriptions taken during his 1850–1856 stay in Tsurai, Trinidad Bay, Northern California, near present-day Eureka. Reproduced in *The Four Ages of Tsurai: A Documentary History of the Indian Village on Trinidad Bay*, edited by Robert Heizer and John Mills, University of California Press, 1952

there were many different arrowhead designs as well.

Many factors go into making an arrow fly well. The length of the arrow, its weight, and the straightness of the shaft are all important. But another important factor is the *fletching*. Fletching refers to the way feathers are chosen, prepared, and placed on the end of an arrow. The way the feathers are cut and attached can make an arrow look beautiful, but the main purpose is to make the arrow fly where it is aimed. The size and shape of the feathers and how they are attached on the arrow shaft are ways of working with the natural laws of physics so the hunter can control the arrow's path.

Different tribes used different methods of fletching. Both the Miwok Indians of Northern California and the Chumash of Southern California used the same side of split feathers. An arrow maker would take three split feathers and attach them to the back of the arrow. The Cherokee did their fletching in a similar way, but often fastened the three half-feathers in slanted positions, adding a slight twist. In the Amazon region and among the Guarani Indians of southern Brazil, the fletching is done with just two half-feathers.

Activity 1:
All About Arrows

This activity is an opportunity to look at several different arrows, study the way they're made, and begin to think about how the design of an arrow affects the way it flies.

Take a close look at your arrows. Notice the materials that each part of the arrow is made from and the

A YOKUTS DEER HUNTER PREPARES TO SHOOT. NOTICE THE THREE
HALF-FEATHERS OF THE ARROW'S FLETCHING.

ACTIVITY 1: ALL ABOUT ARROWS

Is there anything special I should know?

- This activity is recommended for ages 10 and up
- You can do this on your own, or in a group
- If you look at arrows in different places, taking notes and making sketches will help you remember and compare the details

How much time will I need?

- About half an hour, plus whatever time it takes to acquire sample arrows

What materials will I need?

- An assortment of arrows

NOTE: The illustrations in this chapter will be helpful, but you'll learn a lot more if you can examine and handle some real arrows. If there's a museum nearby that has Native American arrows, arrange a visit. Ask if they have information about how the arrows were made or used. You can find modern arrows in some sporting goods stores. (Look in the Yellow Pages under "Archery.") Inexpensive target arrows cost about $2 each. If there is no such store near you, check with physical education departments of local secondary schools or colleges that offer archery and see if you can borrow a few arrows from them.

The Spin

An important concept in flight stability is the way a projectile spins along its long axis. This rotation allows a long projectile to stay in a fixed direction as it moves through the air. It's what makes a great football quarterback's pass travel a long and accurate path.

You can make an arrow spin as the Cherokee and Guarani do by giving a twist to the feather fins. But the Chumash Indians found that using "only the feathers from one side of the bird's tail . . . and always the same side of each feather" caused some spinning. In the case of the Yokuts' arrows, Anne Gayton noted in 1929 that "the feathers were not spiraled but had an almost imperceptible turn."

design of the different parts. As you study the arrows, compare them to the arrows pictured in this book. Notice the ways in which they are alike, and the ways in which they are different. Here are some things to observe and think about:

- What are the shapes of the arrowheads or tips, and what materials are they made of?
- What kinds of feathers were used (are they all real?) and what shapes are they?
- How are the feathers attached to the shaft?
- How long and wide are the arrow shafts and what materials are they made of?
- Which part of the arrow is heavier? (Put the center of the arrow on your finger and see if it balances—if not, which part is heavier?)

Activity 2: Launching Rockets

In this activity, you'll launch model rockets and watch them take flight. Then, by making a simple measuring device called an *inclinometer,* and doing some calculations, you'll be able to figure out how high your rockets fly. If more than one group of people is going to measure the flight of a rocket, remember to make an inclinometer for each group.

> ⚠️
>
> If children are doing this activity, be sure that an adult is present for the rocket launching, preferably someone who has experience with model rockets.

Preparing for Countdown

Begin by making sure you're working safely. Read the instructions for the rocket carefully and be sure you understand them. Work out a plan for where everyone will be at the time of the launch. Decide who will give the countdown (3-2-1-blast off!) so that everyone will be safely located and prepared for the launch. Plan to have everyone say the countdown together.

Before the launch, study the structure of the rocket's body. How does it compare to the arrows you looked at in Activity 1?

Blastoff!

It's time for the countdown. Make sure everyone is in position. Then start the count, launch the rocket, and watch carefully for these things:

- What happens at blastoff?

- When does the rocket speed up and slow down?
- What is the shape of the flight path? Does the wind have an effect on the rocket's path?
- Does the rocket wobble or stay straight? Does it rotate?
- What happens after the rocket reaches its highest point?

If you have enough engines, launch one or two more rockets, and observe them the same way. If the wind kept your first rocket from going straight up, try to make changes to correct for the wind and get a true vertical flight.

Making and Using Your Inclinometer

If you want to measure just how high your rocket flies, you can make an inclinometer. Here are the directions:

1. Tape a drinking straw along one straight edge of the piece of cardboard, as shown below.

2. With a scissors or compass point, make a small hole in the cardboard near the end of the straw.

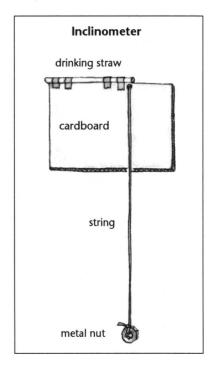

Inclinometer

drinking straw

cardboard

string

metal nut

A MODEL ROCKET LAUNCHING AT AN ELEMENTARY SCHOOL IN SAN FRANCISCO

ACTIVITY 2: LAUNCHING ROCKETS

Is there anything special I should know?

• Before doing this activity, check with your local police or fire department for laws dealing with model-rocket launching. ⚠

• This activity is recommended for ages 12 and up

• You'll need at least three people to do this activity: one to launch the rocket, and two to handle the inclinometer

• Locate a park or athletic field with enough open space for your launch

• Position groups in different locations, all standing about 115 feet from the launch pad

How much time will I need?

• It takes about one hour for the launch, 15–30 minutes to make inclinometers, and 30–45 minutes to do the calculations

What materials will I need?

For the rocket launch:

• Estes Model Rocket launch kit (found at hobby, model, and toy stores for about $35 per kit). Use a rocket that runs on solid fuel pellets and has a battery-driven ignition system that works at a distance. These models have a safety key to prevent accidental launchings. Kits come with two or three engines.

• Batteries for ignition system

• Additional engines (about $2 each) (optional)

To make an inclinometer:

• Measuring tape

• Compass and protractor or other method for making a right angle

• Pencil

• Ruler

• Piece of cardboard, about 8 1/2 x 11 inches or larger

• Plastic drinking straw (not the clear kind)

• Tape or masking tape

• 2-foot length of string

• Scissors or compass point (to poke a hole in the cardboard)

• Heavy metal nut (any size) to act as a weight on a plumb line

3. Pass the string through the hole, and knot it on the back of the cardboard. Tape the knot to the back of the cardboard so it won't slip through the hole.

4. Attach the heavy metal nut to the other end of the string to create a plumb line.

For this part of the activity, you'll need at least three people: one launcher and two measurers. The measurers will position themselves at a distance from the launch pad. You might try 35 meters (about 115 feet).

Measure the distance between the launch pad and the group of measurers and record it. ✋

Each group of measurers has two jobs. The sighter will hold the inclinometer and sight through the straw, keeping the rocket in the sight as it moves up. As soon as the rocket reaches its highest point, the sighter will hold the inclinometer still and call for the taper. The taper will hold the string (plumb line) against the cardboard and tape it in place. If you have a large group, others can be observers, or they can make and use inclinometers of their own.

If you think the wind might affect the flight of your rocket, and you have enough extra engines, do a few practice launches. Work on finding the best angle for the launch to correct for the wind and send the rocket straight up. The sighters can use these trial launches to practice keeping the rocket in sight with the inclinometer.

When the sighters are ready, they can tell the launchers to proceed. When the rocket reaches its maximum height, the sighter should stop moving the inclinometer and hold it still. The taper must immediately hold the plumb line in place against the cardboard and

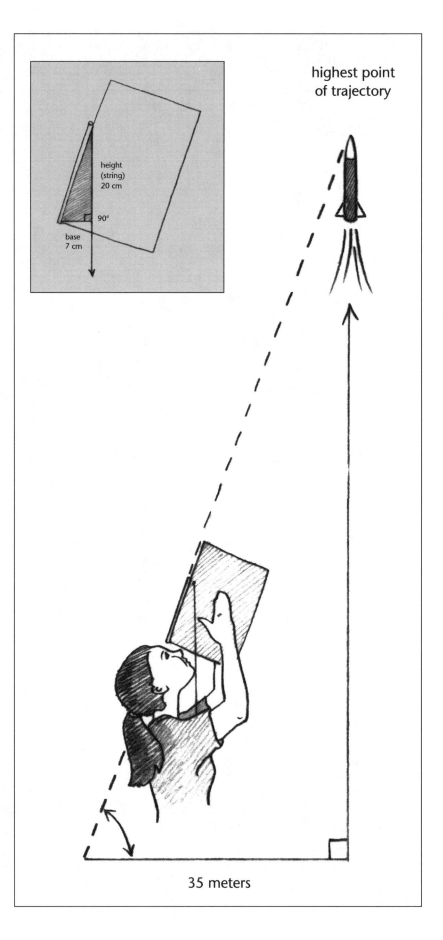

highest point of trajectory

height (string) 20 cm

90°

base 7 cm

35 meters

then tape it down. Once the plumb line is taped down, the taper or an observer should draw a straight pencil line on the cardboard along the plumb line. This records the angle of sighting at the rocket's highest altitude.

Calculating the Rocket's Altitude

Draw a second line, perpendicular to the plumb line, from the plumb line to the bottom of the straw. That line represents the distance on the ground from the launch pad to where the measurers are standing. Now you have a piece of cardboard with a right triangle drawn on it. The side of the cardboard with the straw is the hypotenuse of the triangle.

The triangle on the cardboard is related to another triangle at the launch site. What real points do you think the three points of the cardboard triangle represent? Draw a diagram, filling in the distances that you know.

Think how these triangles are related. They are *similar* right triangles, and if the base of one is twice as long as the base of the other, then you know that the hypotenuse of the larger triangle will be twice as long as the hypotenuse of the smaller triangle. The larger triangle will also be twice the height of the smaller triangle. The two triangles are proportional, and the proportion is 2:1.

PORTABLE ROCKET LAUNCHERS WERE USED TO LAUNCH POISON-TIPPED AND FLAMING ARROWS DURING BATTLES IN FOURTEENTH-CENTURY CHINA.

ACTIVITY 3: FLIGHT STABILITY

Is there anything special I should know?

• This activity is recommended for ages 10 and up

• You can work by yourself or in a small group

How much time will I need?

• About 30–45 minutes

What materials will I need?

• Arrows from Activity 1

• Model rocket(s) from Activity 2

• Yardstick

• Cardboard

• Tape

How can you use similar triangles and the information on your inclinometer to find the height of the rocket's flight?

The larger triangle is made up of three sides: the base is the distance from the launch site to the measurers (on the ground); the hypotenuse is the imaginary line from the measurers to the top of the rocket; and the third side is the height of the rocket's flight. You know the length of the base—the distance along the ground—is 35 meters.

How long is the base of the smaller triangle? Once you know that, you can find the proportion between the two triangles.

Let's say that the actual distance along the ground is 35 meters (m), and the base of the triangle on the cardboard is 7 centimeters (cm) long. That means the proportion of the pencil line to the actual distance is 7 cm to 35 m (which can be written as 3,500 cm). So 7 cm on the cardboard corresponds to 3,500 cm in real life. (See page 138.) The proportion between the cardboard and real-life lines is 7:3,500. To simplify, divide each of those numbers by 7, and you'll get a proportion of 1:500. That means the real distances are 500 times larger than your drawing.

Now use this method to calculate the proportion of the base of the large and small triangles in your rocket launch. 🌢

Once you know the proportion of the base of the smaller triangle to the base of the larger triangle, how can you use that to find the height of the rocket? 🌢

If, for instance, the plumb line side of the smaller triangle measures 20 cm, that means the rocket reached a real height 500 times larger, or 1,000 cm. It went 100 meters into the air.

Measure the plumb line portion of the triangle on your inclinometer. Use the method above to figure out the maximum height that your rocket reached. 🌢

If you worked with several groups of measurers, you may find that you all got different results. If so, compare the drawings that the different groups made and compare their measurements.

Why might it be hard to come up with a single answer to this problem, even if no one made any math errors? Think about the way you measured distances along the ground, and have several people measure the height of the same tree to evaluate the imprecision in the use of inclinometers. 🌢

Activity 3: Flight Stability

Only a carefully designed arrow will fly straight through the air. As an arrow flies, two forces act on it: gravity and air resistance. These two forces work together to keep an arrow flying straight only if the arrow's weight is properly distributed. If the arrow's weight isn't properly distributed, the arrow will wobble or veer off course.

The two keys to proper weight distribution in an arrow are *center of mass* and *center of pressure*.

Center of Mass

The center of mass—the point of balance for any object—is important for flight stability. If you pass a pin through the object's center of mass, you can turn the object in any direction and it will stay in that position—it will be perfectly balanced.

Here's a simple activity you can do to understand center of mass. Hold a model rocket by putting one finger under each end of it. Slide your fingers together toward the middle of the rocket. Notice that it remains balanced. When your fingers meet, you have found the rocket's center of mass.

ACTIVITY 4: BUILDING BETTER ROCKETS

Is there anything special I should know?
- This activity is recommended for ages 10 and up
- You can work by yourself or in a small group

How much time will I need?
- About 1–2 hours

What materials will I need?
- Party balloons
- Arrows and rockets from Activities 1 and 2
- Drinking straws
- Sheets of paper
- Thin cardboard
- Scissors
- Cap from a marking pen
- Masking tape
- Small pieces of paper (self-adhesive notes like Post-Its work well)

How can you be sure that what you have found is the center of mass? Instead of using the rocket, try this with a yardstick. Where do you think your fingers will meet? Is the center of mass for a yardstick in the same position as the center of mass was on the rocket?

With the yardstick, you'll find that your fingers meet right in the middle, at the 18-inch mark. On the rocket, the center of mass was closer to the front.

If you have some arrows available, try the same test with them. Is the center of mass of the arrows at the midpoint, as on the yardstick, or toward the front, as on the rocket? Where would you expect the center of mass to be on a 2-foot board? On a spear?

You have probably discovered that the center of mass is nearer to the front of the arrows. This is true in all successful arrows, rockets, and other flying objects. The center of mass is forward of the midpoint of the object.

Center of Pressure

The center of pressure is the point at which you can hold an object between two fingers and move it through the air without changing the direction in which it's pointing.

To get an idea about center of pressure, first use two fingers to hold a yardstick at its center of mass, then move it through the air. Walk forward, backward, and sideways, keeping the yardstick free to wobble around its center of mass. What do you notice?

You will see that no matter what direction you move in, the yardstick stays pointed in the same direction as when you started. For a yardstick, the center of pressure and center of mass are in the same place. But then, a yardstick isn't a very successful flying object.

Now tape a large cardboard fin to one end of the yardstick (see the illustration below). Hold the yardstick at its center of mass, leaving it free to wobble, and walk forward,

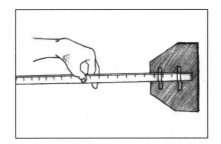

backward, and sideways. Now what do you notice?

The fin changed the location of the yardstick's center of pressure. This time, if the yardstick's long body was pointing in a different direction from the one in which you were walking, the air dragged on the fin. Like a weather vane that lines up in the direction the wind is blowing, the yardstick will soon line up in the direction of travel. The fletching on an arrow acts like the fin. The feathers help keep the arrow going straight.

Experiment with holding the finned yardstick, an arrow, and a model rocket. (Hold the rocket horizontally with one finger on top and one on the bottom. The rocket should be able to move in a horizontal plane.) See if you can find each object's center of pressure and center of mass.

You'll find that the center of pressure is closer to the back of the object—nearer to the fins or feathers—than it is to the midpoint. For any successful flying object to go forward in a straight line, the center of mass has to be in front of the center of pressure.

Activity 4: Building Better Rockets

Think of the things that help an arrow, a rocket, or a dart fly straight and true. All three are long and thin. They all have more mass near the front end. They all have feathers or fins at the rear. Now think of a balloon. Would you expect it to fly like an arrow? ◑

Blow up a balloon and then let it go. What happens? ◑

How could you change the balloon to make it travel in a straight path like a rocket or an arrow? What materials could you use to turn a balloon into a balloon rocket? ◑

You can try out your own designs to build a balloon rocket, or you can use the list of materials here. As you work, you may want to record what you do in words or pictures each time you make a change to your balloon rocket. Then carefully watch how it flies. ◑

One kind of balloon rocket might look something like a Chinese bottle rocket. Stick two or three straws into each other and tape them to the outside of the balloon neck. How did it fly? ◑

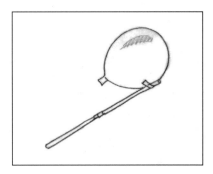

Now try taping the straws to one side of the balloon, as shown above. You will probably notice that this straw "tail" makes your balloon fly straighter. Attaching small paper or cardboard fins to the ends of the straws may improve the flight even more. Try various lengths and numbers of straws, and various sizes and numbers of fins. Blow up your balloon rocket again and see how it flies. ◑

Remember that the center of mass of an arrow is always toward the front. What can you do to make your balloon rocket a little heavier in front? ◑

Try sticking a "nose" on the front of the balloon. A small blob of wet paper towel attached with masking tape will do the job. You can also tape the cap of a marker to the front of the balloon. (To avoid injuries, DO NOT make the nose out of anything sharp or pointed.) ⚠

Experiment with your balloons; try attaching small weights (pennies are good), straws of various lengths, and fins in different locations. The goal is to have your balloon rocket fly straight to where you want it to go. ◑

Rockets Away!

If you're working with friends, you might want to have a balloon rocket tournament. Whose rocket goes the highest? Travels the farthest? Comes closest to hitting a target? What specific features do you think make them each work so well? Are they the ones you thought would fly best? Experiment, and have some fun! ◑

This activity was developed by Maurice Bazin and Modesto Tamez.

Making Connections

- Many people around the world hunt, grow, or gather their food. Where does your food come from? How do you think people in the future will get their food?

- Have you ever used a bow and arrow? Thrown a javelin? Played darts? Where do you see, or use, projectiles in your everyday life?

Recommended Resources

Fadala, Sam. *Traditional Archery.* Mechanicsburg, PA: Stackpole Books, 1999.

Kroeber, Theodora. *Ishi in Two Worlds: A Biography of the Last Wild Indian in North America.* Berkeley: University of California Press, 1988.

Margolin, Malcolm. *The Ohlone Way.* Berkeley: Heyday Books, 1978.

Stine, G. Harry. *Handbook of Model Rocketry.* New York: John Wiley and Sons, 1994.

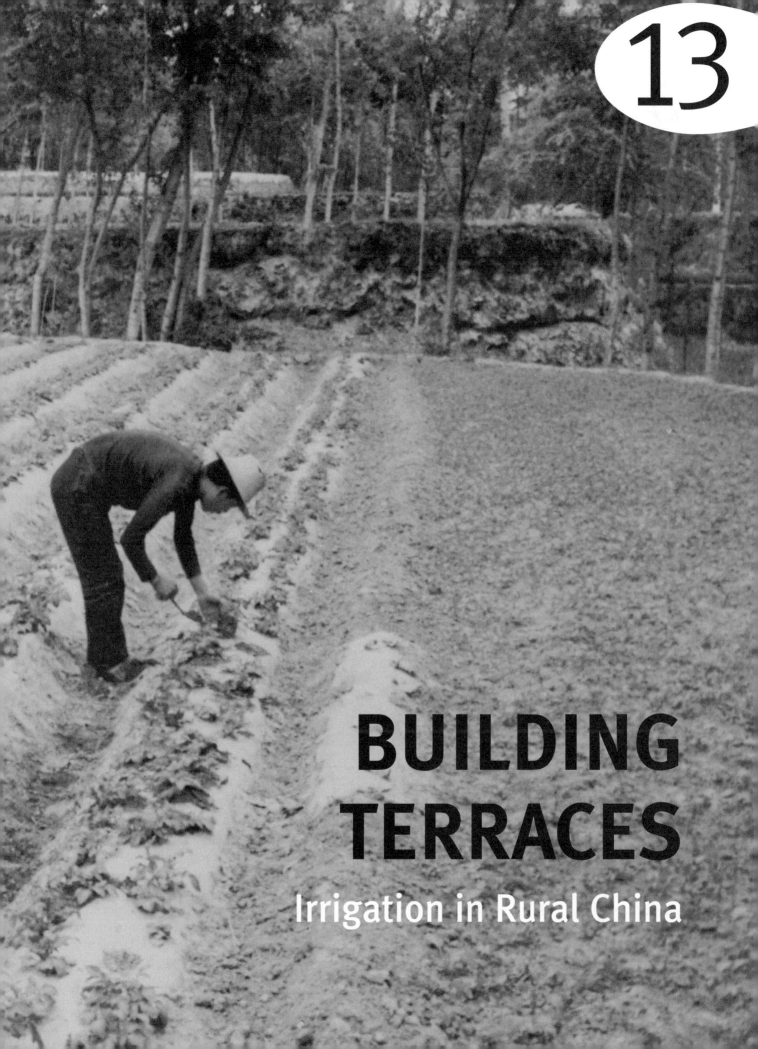

13

BUILDING TERRACES

Irrigation in Rural China

BUILDING TERRACES
Irrigation in Rural China

Suppose your family farmed for a living, but all your land was on hills. How would you create flat fields to grow your crops? How would you stop water from flowing down the hills so you could water your fields? All over the world, farmers solve these problems by building complex irrigation systems that direct water along giant steps, or terraces.

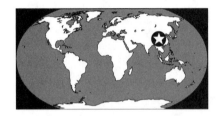

The Tibetan Plateau

The Himalayan mountain range in central Asia is said to be the "roof of the world." To the north of this tallest terrain in the world lies the Tibetan Plateau, an area with an average elevation of more than 11,000 feet (3,500 meters). In the northeast corner of the plateau is a region that Tibetans, the original inhabitants of the area, call *Amdo*, or the "outer northern region." China, which currently has political control over the area, calls it *Qinghai*, or "blue-green sea," because of the large salty lake found there.

Several groups of people live in this area. Among them are Tibetans from the south; Mongols from the north; two branches of Muslims, the Hui and the Sala; the Tu, a mix of Mongol and Tibetan; and the Han, the majority in China.

One of the most striking features of Amdo-Qinghai is its dry, hilly landscape. Because of the high altitude and low rainfall, there is

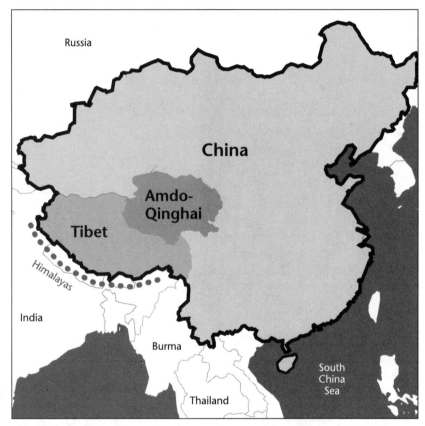

little vegetation. In winter, the hillsides are beige. In summer, a patchwork of small green fields winds down the narrow valleys. On these fields, a year's worth of food grows precariously along the sides of dry hills. The shallow rivers and streams that run along the bottoms of these narrow valleys are usually fed by springs higher up in the hills.

These hillside fields have two amazing features. First, the only water available to them comes from scanty tributaries. Second, the people of the area have made the fields precisely level, even though the natural landcape has no level ground. The ability to level the fields and deliver water to them represents a host of technological achievements that reveal an ancient, elementary genius with which the people of the area obtain food, year after year, from the harsh environment.

The problem of growing crops on hillsides is not limited to Amdo-Qinghai. For thousands of years, people all over the world have altered their parts of the earth in order to produce food on rugged terrain. In much of China, India, Nepal, and Pakistan, the profiles of hills and mountains against the sky is not natural but human-made: a series of steps, like giant staircases, spiral up their sides. You can find the same landscape in the hills around Machu Picchu and other Inca sites in Peru. The steps are terraces, and their flat tops are homes to farmers' fields.

Most fields in Amdo-Qinghai range from two acres down to the size of a small bed. The size of a field

A SECTION OF STEEL TUBING CARRIES IRRIGATION WATER OVER A GULLY.

THIS VALLEY IN THE AMDO-QINGHAI REGION STAYS GREEN BECAUSE OF THE IRRIGATION SYSTEM THAT BRINGS RIVER WATER TO IT. WITHOUT IRRIGATION, THE VALLEY WOULD BE AS BROWN AND DRY AS THE MOUNTAINS ABOVE IT.

WHAT'S IT ALL ABOUT?

There are two activities in this chapter. Each will give you a chance to tackle a problem that challenges mountain farmers throughout the world. As you work, you'll investigate:

- What some of the problems of farming in hilly country are
- What you have to know to build an irrigation ditch that works well
- How flat fields can be built on steep hillsides, and how the fields can be watered
- Why farmers need to work cooperatively

Activity 1: Building Irrigation Ditches

In this activity, you'll create a miniature irrigation canal to take water from a high place to a lower place. The challenge is to lose as little dirt to erosion as possible as you move the water where you want it.

Activity 2: Building Terraces

In this activity you'll create miniature fields on a sloping surface, just as mountain farmers have done for centuries. The challenge here is to make fields that are flat enough to hold irrigation water.

is determined in large part by the hill's steepness and the shape of the natural terrain.

To make each field perfectly level and perfectly flat, a farmer uses his eyes, a shovel, and a two-wheeled cart with a long beam that scrapes the ground to flatten it. A ridge around the edge of each field makes it into a shallow reservoir several inches deep. When the fields are filled with water, the water stands for days while the crops drink it in. The fields are irrigated three or four times each summer.

The irrigation ditches that bring the water to the fields seem quite simple at first. They are made of the same dirt that's in the fields, and the water flows through the ditch on its way down the hill to the field. But the situation is a little more complex. In Zhao Jia Zhuang, the village where most of the research for this chapter was done, the spring that feeds water to the fields is almost two miles (three kilometers) away, across extremely hilly terrain.

The ditch through which the wa-

ter runs must follow every turn and twist in the hill, taking the water down little by little, at just the right speed. If the water goes too slowly, it will overflow the ditch. If it goes too fast, it will cut away the dirt and break out of the ditch. The planning and engineering required to make this system work is astonishing.

Along the two miles of ditch, there's only about 300 feet of steel tubing or rock and concrete reinforcements. The tubing carries the water over gullies where other seasonal streams flow straight down the hill (see page 147, top photo). The rock and concrete help protect the ditch along the steepest sections, where the water flows the fastest.

The ditch carries water to hundreds of fields in several villages. To irrigate his fields, a farmer chops a hole in the wall of the surrounding ditch and shovels the clump of dirt from the wall into the middle of the ditch, downstream from the hole, so the water is forced to flow into the field instead of continuing downstream. When the field is full (usu-

ally a matter of minutes), the farmer replaces the clump of dirt in the wall of the ditch. Then the water flows down to the next field.

This technique is simple, but amazingly effective. If you could look down from above, you would see that the irrigation ditch doesn't look like a snake, it looks more like a tree with dozens of branches. The staircase of fields is actually a spiral, roughly following the sloping contour of one branch of the irrigation ditch as it wraps around the hill. Each field in this spiral staircase is only a few inches below the previous one. But if you followed a line going straight up the hill, you would have to step up several meters from one level of the staircase to the next.

The soil in Amdo-Qinghai looks inadequate for growing anything. It's pale brown or yellow in color and has extremely small, claylike particles. This material, called *loess* (rocks ground up by glaciers), seems better suited to making pottery than to growing crops. But it is fairly stable in a vertical face. The courtyards and barns in Amdo-Qinghai are made from loess, packed into walls sometimes 15 feet (4 $^1/_2$ meters) tall and 3 feet (1 meter) thick at the base. The walls are stable, even though they have no other material for structural support, and can last for decades. The same is true for the vertical walls between terraces.

At the edge of many fields, farmers plant rows of trees to help support the vertical part of the terrace. In most places, trees also line the irrigation ditches: Their roots help keep the walls of the ditch from being washed away, and water flowing through the ditch keeps the trees alive. The trees are skinny and straight, pruned to be branch-free, and are used for beams in the buildings of the village.

Activity 1: Building Irrigation Ditches

Imagine that you are a farmer. Your water comes from a spring at the

TO DIVERT THE FLOW OF WATER INTO A FIELD DURING THE MIDDLE OF THE NIGHT, A CHUNK OF WALL WAS CUT OUT AND PLACED IN THE CENTER OF THIS DITCH.

top of a hill, and your fields are downhill from the spring. The challenge of this activity is to build an irrigation ditch to carry water down the hill to your field. You'll build a model field inside a box, then pour water into the top of the ditch. Your

ditch should carry the water down to the bottom of the box.

Follow these directions to build your model irrigation ditch:

1. If necessary, trim your box to a depth of about eight inches.

ACTIVITY 1: BUILDING IRRIGATION DITCHES

Is there anything special I should know?

- This activity is recommended for ages 10 and up
- You can do this alone or with a friend
- Note that if the dirt you're using is loose and dry, it tends to float. The flowing water can then carry a "blob" of water and dirt that could break the side of your ditch

How much time will I need?

- About 2 hours

What materials will I need?

- Large, shallow, sturdy box 5–8" deep and at least 12" wide and 20" long (wax-lined banana boxes from the grocery store work well)
- Large plastic garbage bags
- Dirt (not potting soil)
- Small shovel or trowel
- Adhesive tape (optional)
- Large and small boards for props
- Bucket of water
- Some twigs, leaves, and rocks
- Pitcher
- Funnel

THIS DIAGRAM SHOWS THREE IRRIGATION DITCHES BRANCHING TO BRING WATER TO SIX TERRACES.

TREES AND IRRIGATION DITCHES WORK TOGETHER: TREE ROOTS HELP KEEP THE WALLS OF THE DITCH FROM ERODING, WHILE THE WATER IN THE DITCH NOURISHES THE TREES.

2. Line box with two layers of garbage bags. If the bags are small, cut them open and spread them over the box. Tape them together if you need to. The plastic liner should drape over the sides of the box.

3. Fill the box with about five inches of dirt and pack it down. You'll want to wet the dirt to help pack it down, but don't make it too muddy.

4. Re-create the slope of the hillside by propping up one of the short ends of the box on a wooden block or log. If the bottom of the box sags from the weight of the dirt, put a large board under it or put several props under it at different places.

See if you can build a ditch that will meet these two challenges:

• You want the maximum amount of water to reach the field.

• You want a minimum amount of dirt to be washed away by the water.

You can make your ditch in any shape and any width you like, as long as it goes down from the top of the box to the bottom. ●

Testing Your Model

When your ditch is finished, pour water from the pitcher through the funnel into the top of the ditch. Your model is successful if it brings the water down the slope without bringing along much dirt. If you're not satisfied with the results, see if you can figure out what needs to be changed. Then repack the dirt in your box and make a new ditch. ●

You might try changing the shape of your channel (make it curve more, or make turns less tight). You could add some rocks and sticks to help it keep its shape, organize the flow of water to pre-

vent erosion, or make the ditch wider or narrower. Now test your ditch again. Were your results different? ✊

You probably noticed that if the water is allowed to run straight downhill, it will start to erode the sides of the ditch. After a while, the water will dig a deep canyon in the hill. If the ditch is angled or curved but its slope is too steep, water will flow over the edge of the ditch at

THE MAIN TRUNK OF AN IRRIGATION DITCH

Farming by Moonlight

Throughout the growing season in Amdo–Qinghai, water is constantly being distributed to the fields. One moonless night when I was visiting the village of Zhao Jia Zhuang, the water was coming down for a certain line of fields at eleven o'clock in the evening. All across the hillside, cigarettes were glowing, but very few flashlights. Farmers called to each other and occasionally cursed about a broken section of the ditch.

The farmers were there for hours: first waiting for the water, then directing it onto their fields, then patching up the ditch so it would not destroy their fields while taking water to the fields farther down. Finally, they made sure the entire perimeter of their field was tight, holding in the precious water for the crops to use.

The residents of Zhao Jia Zhuang told me that irrigating is the most laborious work of the entire season, even harder than planting or harvesting, because of the hours. There is an elaborate schedule for irrigating the fields in the village. It is not a schedule with specific times, but rather an order in which fields will get water. There is a lot of debate and discussion about this order, and the people in charge of it hold much power in the village.

Even when the order is worked out, a family still doesn't know when the water will come to their field. Each must wait for news that the people before them on the schedule are getting water and then prepare to receive the water themselves. It may happen at noon or at three o'clock in the morning. It is very much an "every man for himself" situation: If you are not there when the water comes to your field, no one is going to direct it onto your field for you.

—Curt Gabrielson

ACTIVITY 2: BUILDING TERRACES

Is there anything special I should know?
- This activity is recommended for ages 10 and up
- You can do this alone or with a friend

How much time will I need?
- About 1 hour

What materials will I need?
- Box with dirt from Activity 1
- Extra dirt (moist but not muddy)
- Boards and wood to prop up box
- Small shovel or trowel
- Ruler
- Seeds for tiny plants such as grass or cilantro (optional)

the first turn. If the water flows too slowly around a wide bend, you may notice that a lot of dirt is deposited on the inside of the bend.

Activity 2: Building Terraces

Imagine that this is your land, where you must plant the crops that will support your family for the next year. To do this, you'll need to create fields that are flat and level. Each field must also be able to hold irrigation water.

The terraces used in most of the world look like giant, flat-topped staircases. The fields are the flat, level parts of the steps; vertical walls separate one field from another.

You can make terraces in the same box you used for the irrigation ditches. Dump out any extra water and pack the dirt down flat again. Tilt the box lengthwise, at a slope of between 15 and 45 degrees (see photo above). Then, if you want, you can follow these steps, or you can create a step pattern of your own:

1. Start at the top of the box. To mold the highest terrace, push against the dirt with the side of your hand until you've made a small vertical wall, about an inch high, all the way across the box.

2. Now use the palm of your hand to pat down the dirt on top of the wall. You should end up with a flat surface (about 4 inches wide) that's at a right angle to your little wall.

3. Add dirt if you need to. Use your hand to tamp down the dirt on the top (flat) part, and use the ruler to push in on the vertical wall and to make a sharper angle.

4. Continue this process, moving down the box. You should be able to make at least four terraces. You may have to remove some dirt to make the lowest terrace.

5. When your terraces are done, use the extra moist dirt to form a small ridge around the sides and front of each terrace. This will help keep irrigation water in your tiny fields.

Irrigating the Fields

In real terraces, fields that are side by side will be in line for water from one ditch. A field that is directly

above another one will be watered by a different irrigation ditch.

You don't have to build a ditch system that will get water to flow from one terrace down to the next one. That would be complicated, and your model isn't quite big enough to make enough interconnecting ditches. Instead, pour some water into each field so that it forms a shallow pool held in by the ridge around the edge.

Look at your result. Does the water cover the fields evenly? Are there places where the dirt sticks up above the water? If this were a real field, crops in those areas wouldn't be getting enough water and would not do well.

The model area of land you're working with is a rectangle, but in mountain areas, the terraced fields aren't so square or even. Usually, a hillside slopes irregularly and also wraps around a mountain, so the land is divided by rocks, gullies, and other types of terrain. In many parts of the world, the lower slopes of

The Necessities of Life in Amdo-Qinghai

I learned a lot about food and the necessities of life when I was in China. Most of my students in Amdo-Qinghai were from villages, and I took every opportunity to go to their homes and get a closer look at rural life. I was struck by how meager their kitchens seemed—yet meal after wonderful meal came out of them. Each ingredient is precious and cared for with an attitude rarely found in the West, or even in the cities in China—a rational response to the very real hardships of acquiring food in the first place.

Wheat is the staple in Amdo-Qinghai, and each grain of wheat is cut from the field, shucked, hauled to the threshing ground, beaten out of its hull, winnowed, gathered up, sorted and bagged, then stored carefully in a bin in someone's home. Often the bin is in the bedroom, because the bedroom is one of the cleanest rooms in the house.

Each month or so, the wheat is taken out of the bin and spread out on the porch to dry and be picked clean of bugs. When it comes time to eat the wheat, it is ground into flour, kneaded, rolled out and cut for noodles, or made into biscuit-like pieces for cooking.

Of all these operations, only the threshing and the grinding are done by machine, and these machines have been implemented only recently. The complete absence of any form of waste is then quite natural and understandable. These people are living close to the edge of subsistence, and it would be foolish to throw away anything that might help keep them alive. It is no different with their soil and water. Every bit of land is used for growing food, and the water flowing from the springs is used for irrigation twenty-four hours a day.

—Curt Gabrielson

mountains are cluttered with rocks and boulders that are brought down by water and gravity. Farmers have to work around all these areas to irrigate their terraces.

To improve the flatness of the fields, you can scrape off some of the dirt in higher places and spread it in the lower parts. That's just what farmers do in terraced fields. The vertical walls are always in the process of slowly crumbling down onto the fields below them, and the irrigation water continually carries more silt onto the fields. The farmers take this "new" dirt and spread it evenly over their fields to make them flat once more.

Once you've made any needed repairs, try adding water again to see if your terraces are flatter and more even.

You probably noticed that every time water is poured into your model fields, the terrain changes a little. This is certainly true in real terraced fields. The entire mountainside is in a constant state of change, so the work of maintaining the terraces is never done.

Planting Crops

If you want, you can try planting your miniature fields. Find a place to keep your terraced box where it won't be disturbed, and where it will get some sun. Then plant seeds or seedlings in your terraced fields and water them regularly. Seeds for small plants like grass or cilantro will probably work best on your tiny terraces.

Better yet, try this in a real garden, taking advantage of the terrace-building technology you've been experimenting with.

This activity was developed by Curt Gabrielson.

Making Connections

- What are the most important resources where you live? How do people share them?

- What role does cooperation play in maintaining and irrigating the terraces of Amdo-Qinghai? What would happen if the farmers did not cooperate?

- How is cooperation important in your life? What problems have you had with cooperation in different settings?

- You probably turn on a faucet to get water in your house. But where did that water come from? A well? A water tank? A reservoir? See if you can trace the origins of the water you use every day.

Recommended Resources

Feigon, Lee. *Demystifying Tibet*. Chicago: Ivan R. Dee, Inc., 1998.

Gray, Donald H. *Biotechnical and Soil Bioengineering Slope Stabilization: A Practical Guide for Erosion Control*. New York: John Wiley and Sons, 1996.

Zhang, Mingtao. *The Roof of the World: Exploring the Mysteries of the Qinghai-Tibet Plateau*. New York: Harry N. Abrams, 1982.

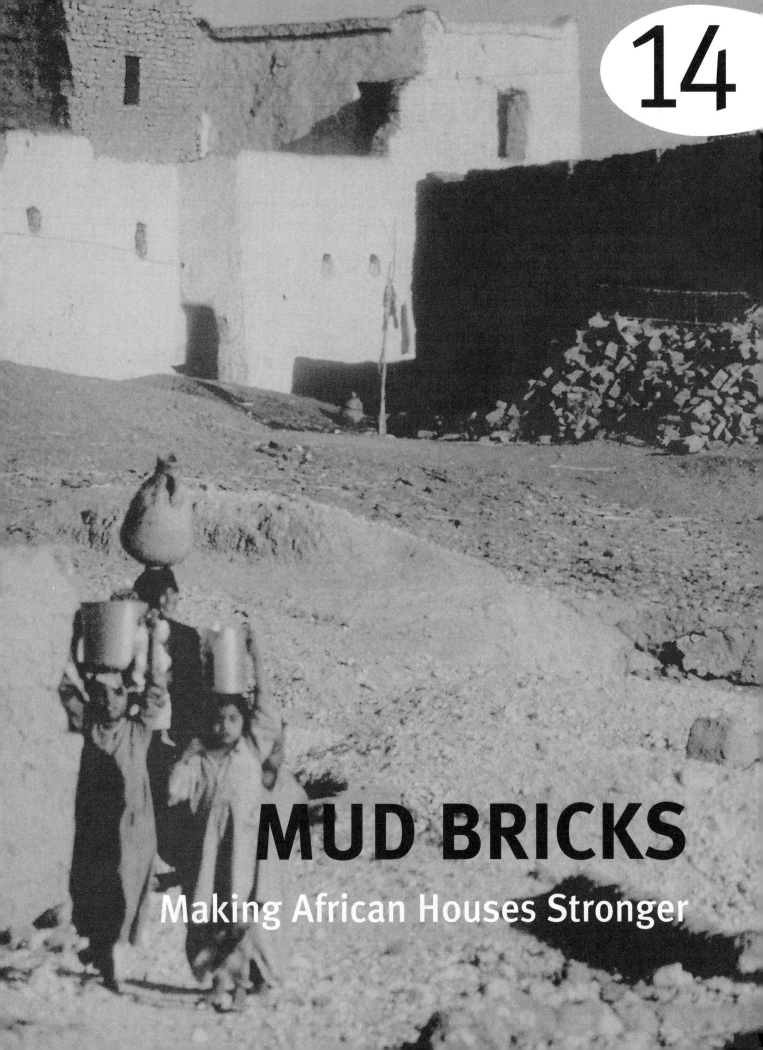

14

MUD BRICKS

Making African Houses Stronger

MUD BRICKS
Making African Houses Stronger

All over the world, people who build their own houses apply math and science to figure out how to make strong, long-lasting structures. In this activity, you'll do what a structural engineer does: test materials and decide how suitable they are for building.

African Mud Housing

In many parts of the world, including large parts of rural Africa, people make their houses themselves or hire a person from their village to build it. The materials used to build the houses—mostly stones, poles, mud, reeds, and leaves—come from nearby places.

Although these materials are easily available, they create structures that need a lot of attention. Rather than spending time constantly repairing their houses, individuals and families might be able to use that time in better ways—taking care of children or elders, tending animals, going to the market, gathering wood and water, or taking a paying job.

Suppose you were responsible for finding ways to make houses that needed less maintenance, but you knew that there wasn't much money available. You might consider the use of a building material called soil-cement.

Cement is a mixture of clay, limestone, and metal traces that

WORKERS IN NORTH AFRICA USE MUD BRICKS
TO RESTORE THE UNDERGROUND CHAMBER OF A MOSQUE.

WHAT'S IT ALL ABOUT?

By making miniature bricks out of varying proportions of cement and dirt, you will begin to think about some of the issues people face when they build houses in rural Africa and other parts of the world. You'll investigate:

- What makes mud brick a good building material
- How to test your sample bricks for strength, resistance to erosion, and other factors
- How to be sure that your tests are reliable and repeatable

Is there anything special I should know?

- This activity is recommended for ages 12 and up
- It's best to work in small groups (the materials listed below are enough for a whole group)

How much time will I need?

- You'll need to do this activity in four steps, over a period of many days (the days don't have to be consecutive)

Step 1: Make a variety of miniature soil-cement bricks, 1–2 hours

Step 2: Let the bricks set and harden, 2–3 days

Step 3: Test the bricks for various stress-resistant properties: there are 4 tests; allow at least 2 hours per test over a period of at least 2 days (some may need to run overnight)

Step 4: Organize the results of your tests, 1–2 hours

What materials will I need?

For making the molds:

- 8 to 12 small cardboard boxes, at least 1 inch deep (jewelry, greeting card, diskette boxes)
- Lightweight cardboard for making partitions

For making the bricks:

- Soil (enough to fill a 1- or 2-gallon bucket). NOTE: All you need is a bag of plain dirt. If you use soil from your backyard, pick out any rocks, sticks, or other unnecessary creatures or objects. If you buy a bag of soil, make sure it contains no vermiculite or other additives.
- Cement (enough to fill a 1-gallon bucket). NOTE: You can buy a 100-pound bag of Portland cement from a hardware store for a few dollars. If possible, see if you can obtain a smaller quantity. If you know someone who works with cement, ask if they'll give you a 1- or 2-gallon paint-can full. Do not buy mortar, which some hardware stores may call cement. It has sand mixed in and won't work in this activity.
- Sealed container for the cement
- Water
- Plastic measuring cups
- Scales (optional)
- Paper cups
- Large mixing bowl
- Pitcher
- Large wooden spoon, plastic spoon, or spatula

For testing the bricks:

- A pencil or small dowel
- Bathroom scale
- Sandpaper
- Water

has been heated to an extremely high temperature and then ground into a powder. Soil-cement is a mixture of cement and other materials. It's used in many places to improve those parts of a house where soil, dirt, rocks, or mud are normally used by themselves. If a house were constructed of dried mud bricks, for instance, the bricks could be strengthened by mixing cement in with the mud. If the house were made mostly of stones layered with an outer coat of mud, cement could be added to the mud to make the coating more durable.

Mixing cement with soil costs more and takes more time and effort, but it can sometimes make enough of an improvement in a house to be worth the extra money and labor. But because the strength of soil-cement can vary with the quality of the local soil, its added value can also vary. Families need to determine what ratio of soil-to-cement works best in their local conditions, and if, in fact, there is any value to using soil-cement at all.

The Big Question— and Some Smaller Ones

The mud bricks used in rural areas of Africa and around the world are quick and easy to make, but they're not very strong. In this activity, you'll try to find out if adding cement to soil will strengthen mud bricks and, if so, whether it adds enough strength to be worth the added time and expense.

In order to answer these questions, you'll have to think about what makes a "good" brick, how its quality can be tested, how its strength can be measured, and how many bricks you need to test to be sure your results are reliable. It's important to keep good records for this activity. You may want to use a clipboard or notebook so you can write down all your information in one place.

Planning Your Experiment

You'll be testing your bricks to see how well they resist crushing, breaking, abrasion, and erosion. For each test, you'll want to see if the cement really improves the bricks.

You'll need to make bricks with four different soil-to-cement ratios. To make sure your tests are reliable, you'll want at least five bricks of each ratio for each of the four tests. That means you'll need to make at least 80 bricks: 4 different ratios times 5 bricks of each ratio times 4 tests. It's probably a good idea to make 5 or 10 extra bricks, as well. The process of making the bricks is easy, and it's better to have more bricks than you need than to run out in the middle of a test.

The first thing you need to decide is what ratios of cement-to-soil to use. You can make one set of

bricks out of pure mud, but what proportions should you use for the other three sets?

You might think first of a large proportion, like half-mud, half-cement. But what if a small amount of cement makes a big difference, and you never discovered that? You'll also find that dealing in increments of 5 percent and 10 percent will make your calculations much easier.

Here's one suggestion for four sets of soil-cement bricks:

- Bricks of 100 percent soil (and water, of course)
- Bricks with 10 percent cement, 90 percent soil
- Bricks with 20 percent cement, 80 percent soil
- Bricks with 40 percent cement, 60 percent soil

You might also try proportions that are easy to measure, like one-quarter cement to three-quarters soil.

Cement is not a poisonous chemical, but it is a powder, so don't inhale it and don't let it get into your eyes or mouth. Cement irritates some people's skin, so handle it with care. Keep the cement in a sealed container. Close it tightly and put it aside when it's not in use. If some cement powder spills, first pick up the powder and then rinse your hands and the work surfaces with lots of water.

You can measure the soil and cement either by volume or by weight. If you measure by volume,

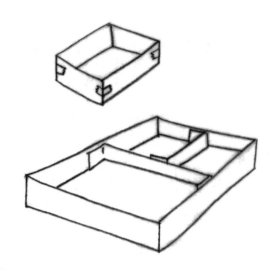

use 1/2 cup cement and 4 1/2 cups of soil (a proportion of 10 percent cement to 90 percent soil). If you have a scale that measures small amounts (like a lab scale or a kitchen scale), you can weigh out 2 ounces of cement and 18 ounces of soil to create the same proportion.

Although the ratios in your bricks will be different, your mixtures may look about the same, so it's important to label every batch carefully.

Step 1: Making the Bricks

You'll be making small, sample bricks, about 2" x 1" x 1". Making 80 full-size bricks takes a lot of cement and soil, and these smaller bricks are much easier to test. When you're testing resistance to crushing, for instance, the force you need to crush the brick will be proportional to the brick's area. With a small brick, you won't need a very large force.

Make the molds by taping strips of lightweight cardboard inside boxes to create partitions. A standard greeting-card box is about 4" x 6" (24 square inches), which will make twelve 1" x 2" bricks. Smaller boxes,

THIS HOUSE IN A TANZANIAN VILLAGE HAS MUD-BRICK WALLS.

such as ones used for earrings, would probably be about 2" x 2" (4 square inches), and will make two bricks. You can use a variety of different-sized boxes, as long as they are divided so the bricks are all about the same size.

Making same-size molds from different-size boxes can be a challenge. You might want to take some

time to plan and make sketches before you start to cut partitions. Then prepare the molds, making sure you have enough to make at least 80 bricks.

If you don't have enough boxes, you can design your own molds by folding, cutting, stapling, or taping sheets and strips of cardboard. You can also use small paper cups as

molds. This will make round "bricks," which may react differently to stress tests than rectangular bricks do.

You've decided on your proportions, and you've made your molds. Now it's time to make bricks.

Start with one ratio of soil-to-cement. Measure out the proportions of each material, either by

THESE MUD BRICKS IN TANZANIA HAVE BEEN SET OUT TO DRY IN A SINGLE LAYER. LATER, THEY WILL BE STACKED LIKE THE BRICKS IN THE BACKGROUND.

weight or by volume. Using a spatula, mix the dry ingredients together in a bowl or other container. Then add some water.

How much water should you use? Experiment by adding water little by little to see what happens. You'll soon discover that there is a minimum amount you need to wet the whole mixture, and also a maximum amount you can add before the extra water runs off. These two extremes are close to each other, so

always add the water slowly. You'll know you have the right amount of water when the mixture becomes a paste.

Carefully spoon the paste into your molds, leveling off the top surface with a knife or the edge of a piece of cardboard. Be sure to label every mold with its soil-to-cement ratio.

Repeat this procedure for the other three soil-to-cement ratios.

Step 2:
Letting the Bricks Set

When cement sets, or "cures," it is not drying out. The water, or at least some of it, is incorporated into new chemical bonds that give hardened cement its strength and bonding power. If you have ever worked with cement or watched people using it, you might have seen that to get a good result, you need to keep the cement damp while it sets. This keeps the water that is needed in the chemical bonds from escaping.

It will take at least 2–3 days for your bricks to set. You may want to:

• put them in a hot place
• put them in a cold place
• put them in the sun
• put them in the shade
• put them in sealed plastic bags

Whatever you decide to do, make sure you use the same method for all 80 bricks.

When you think the bricks have set, try taking one out of the mold. If it falls apart when you handle it, there are two possibilities: The brick may not be completely set yet, or it may just be a really crumbly brick. How could you investigate that problem? 🔵

The best thing to do would be to wait another day, then take another brick out of the mold. If it's still crumbly, you may have discovered that a particular soil-to-cement ratio doesn't make very strong bricks.

Step 3: Testing the Bricks

There are several properties that a brick should have if it's going to be a useful building material:

• **Resistance to crushing:** The weight of the house bears down on the material that makes up the walls, so it's important for bricks to have a high resistance to crushing.

• **Resistance to breaking:** Some bricks might be over doorways or windows, where weight presses down from above with no support underneath. Those bricks have to be resistant to breaking the way a cookie breaks in half.

• **Resistance to erosion:** Water might run down the walls of a house during a storm, so it's important for bricks to have resistance to erosion from moving water. The walls of a house might stay wet for a while during and after storms, so you may also want to test your bricks for resistance to crushing or breaking after they have been soaking in water for some time.

• **Resistance to abrasion:** In windy climates, sand and soil may be blown against the walls of a house. Solid things like trees and shrubs might also rub against the bricks, so resistance to abrasion is also important.

Before you start, you may want to think about the testing process— the tools you'll need, ways you can measure and track your results, and the number of times you should repeat a test to make sure your results are valid.

Test A: Resistance to Crushing

For the first two tests, you will be doing "destructive testing," in which you measure the force or effort it takes to break something. Increase the load or stress gradually, so you can measure the stress just before the brick fails.

To test for resistance to crushing, you need to apply force evenly on the face of a brick that is lying on a flat surface. How could you apply that kind of force? How would you measure it? 🔵

In this test, you will need to record the amount of weight the brick was supporting just before it was crushed. You can do this by adding weight until the brick is crushed. You might want to try piling one-pound weights on the brick,

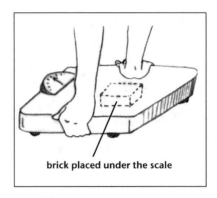

brick placed under the scale

one by one, or put a large container on top of the brick and add water, one pint at a time.

These methods will work fine if you can build up enough weight to crush the brick. But if you put the brick on a table or on the floor and tried to pile enough things on top of it to crush it, you probably found that it was hard to find things heavy enough to do the job.

The simplest technique is to put a bathroom scale on top of the brick, then push down on the scale. The reading on the scale as you press down will be the amount of force you are applying. Be sure to have someone reading the scale who will call out the weight just as the brick is crushed.

Use any method you choose to test your bricks for their resistance to crushing, but remember to use the same method for all the bricks you test and to record your results for each set of soil-to-cement ratios. 🔵

Test B: Resistance to Breaking

You will probably find that the forces you need to break a brick are much smaller than the forces you needed to crush one.

One way to measure the force it takes to break a brick is to put a pencil on a scale, place the brick across the pencil, and push down on both ends of the brick until it breaks. Have someone read the scale and announce the weight that showed just as the brick broke.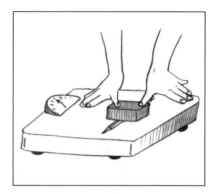

Use the same method for all the bricks you test and record your results for each set of soil-to-cement ratios.

Test C: Resistance to Erosion

You can have fun coming up with ways to test this strength. For the more stable brick, you may need to create a long-term experiment. You could place a brick in the sink, for instance, and adjust the water so that it drips slowly onto the brick overnight. Or you may want to invent a device, like a punctured plastic bag that will drip water onto a brick drop by drop. Or you could go outside and use a hose to create a steady trickle of water that runs over the brick overnight.

Figuring out how to measure the effects of erosion is another challenge. One way is to measure changes in the size of the eroded bricks. Another way is to measure changes in the weight of the bricks. If you use the second method, remember that you have to account for the weight of any water that the brick absorbs.

If you have enough bricks, you could subject one brick from each group to erosion, and then test it for crushing or breaking.

MOST LARGE BUILDINGS IN CITIES TODAY DEPEND ON CONCRETE FOR THEIR STRENGTH. CONCRETE IS A CAREFULLY PROPORTIONED MIXTURE OF CEMENT (FOR BONDING) AND THE PROPER AMOUNTS OF STONE, SAND, AND GRAVEL (FOR BULK AND WEIGHT). MOST OF WHAT YOU SEE IN A CONCRETE WALL IS THE STONE, SAND, AND GRAVEL. OFTEN CONCRETE IS MADE EVEN STRONGER BY PUTTING METAL RODS IN IT. REINFORCING CONCRETE WITH METAL IS NECESSARY FOR VERY TALL BUILDINGS AND BRIDGES, WHICH ARE STRESSED BY TREMENDOUS FORCES.

Test D: Resistance to Abrasion

When things rub together, the resulting abrasion can wear them down. For this test, you have to figure out how abrasion affects your bricks. You also have to figure out a way to measure these effects. You might, for instance, glue or tape some sandpaper to a flat surface. Then you could rub a brick over it repeatedly and see how much of the brick material is gone after a given number of strokes.

A good test has to be repeatable, so you need to be sure you're testing every brick in exactly the same way. There are lots of variables for you to consider:

• How much pressure did you apply when you rubbed the brick on the sandpaper? How can you measure that?

• How long was your stroke? Can you repeat it exactly?

• Was the sandpaper gradually getting smoother (and less effective) because it was wearing out or filling with dirt?

You will need to find ways to solve these problems and standardize your tests. For example, you might put a weight on the brick and use your hand to move the weighted brick back and forth without applying any additional downward force. You could compare how quickly the bricks wear away when you change the sandpaper frequently with what happens when you use the same piece of sandpaper for a long time.

Step 4: Analyzing Your Data

After you've done all these tests, you'll have lots of data about how your bricks reacted to various stresses. How can you organize all that information so that you can use or share it? Think about a good way to show the data for each test. Will a table of numbers show your results? Would the information be more clear on a graph? Did some bricks react well to one kind of stress but not to another? Does one soil-to-cement ratio stand out as being better than others for making strong, durable bricks?

Once you know what ratio of soil-to-cement makes the strongest brick, you can apply that knowledge to a practical question: If cement adds strength to mud bricks, does it add enough strength to make it worth the cost of buying the cement?

You'll have to figure out many things to answer that question. How big is a real mud brick? How big is an average mud-brick house? How many bricks would it take to build one? How much mud and cement would you need for each brick? If your results showed that adding about 20 percent cement to the mud makes the strongest bricks, how much cement would you need to build a house?

You also have to consider the cost of cement and the value of people's time in the place where the cement will be used.

Let's say a bag of cement costs $20 in an African village. A single bag of cement may make one wall of a house quite a bit stronger. But how many days or weeks must a rural farmer work to earn $20?

Answering that question will probably require some research on the economy of a region, and the value that its people place on different kinds of work. In Africa, women do much of the work of building and repairing houses, and the value of men's and women's labor may be considered differently.

As you can see, thinking like an engineer isn't just a matter of doing math and experiments and building things. It's also thinking about people—how they live, where they live, and how they'll use the things that are built.

This activity was developed by Robert Lange.

Making Connections

- What factors affect the cost of a house? Besides the materials themselves, think about what it costs to get the materials to the building site, and what it costs to pay someone to build with them.

- What materials are used for houses where you live? Are they practical choices for your area?

- If you lived in the Arctic, what would you have to consider if you wanted to build a house? What if you lived in a jungle or on a prairie?

Recommended Resources

Dethier, Jean. *Down to Earth*. Translated by Ruth Eaton. New York: Facts on File, 1983.

Elleh, Nnamdi. *African Architecture*. New York: McGraw-Hill, 1996.

McHenry, Paul Graham. *The Adobe Story: A Global Treasure*. Albuquerque: University of New Mexico Press, 1998.

Smith, R. C., and C. K. Andres. *Materials of Construction*. New York: McGraw-Hill, 1966.

NATIONAL RESEARCH COUNCIL SCIENCE CONTENT STANDARDS

	Unifying Concepts and Processes			Science as Inquiry	
Grades	4–12	4–12	4–12	4–12	4–12
	Constancy, change, and measurement	Evidence, models, and explanation	Form and function	Abilities necessary to do scientific inquiry	Understanding about scientific inquiry
1. Sona					
2. Cuica		X	X		
3. Madagascar Solitaire					
4. Quipus					
5. Counting Like an Egyptian	X	X		X	X
6. Breaking the Mayan Code		X		X	X
7. The Mayan Calendar Round					
8. Weaving Baskets					
9. Dyeing		X		X	
10. Tea and Temperature					
11. Rain in a Bottle	X	X		X	
12. From Arrows to Rockets	X	X	X	X	
13. Building Terraces			X	X	
14. Mud Bricks	X	X		X	X

Physical Science					Earth and Space Science	Science and Technology		History and Nature of Science	
4	5–8	5–8	5–8	9–12	5–8	4–12	4–12	4–12	5–8
Properties of objects and materials	Properties and changes of properties in matter	Motions and forces	Transfer of energy	Chemical reactions	Earth in the solar system	Abilities of technological design	Understanding about science and technology	Science as human endeavor	History of science
			X						
X						X	X	X	X
X									
		X			X			X	X
					X			X	X
X									
X	X			X		X	X	X	X
X	X		X				X		
X	X		X			X	X	X	X
X		X				X	X	X	X
X						X	X		X
X	X	X				X	X	X	X

169

NATIONAL COUNCIL OF TEACHERS OF MATHEMATICS MATH CONTENT STANDARDS

Grades 4–12	Number and Operations			Algebra	
	Understand numbers, ways of representing numbers, relationships among numbers, and number systems	Understand meanings of operations and how they relate to one another	Compute fluently and make reasonable estimates	Understand patterns, relations, and functions	Analyze change in various contexts
1. Sona	X	X	X	X	X
2. Cuica					
3. Madagascar Solitaire					
4. Quipus	X	X	X	X	
5. Counting Like an Egyptian	X	X	X	X	
6. Breaking the Mayan Code	X	X	X	X	
7. The Mayan Calendar Round	X	X	X		
8. Weaving Baskets					
9. Dyeing					
10. Tea and Temperature					
11. Rain in a Bottle	X				
12. From Arrows to Rockets	X	X	X		
13. Building Terraces					
14. Mud Bricks	X		X		

Geometry		Measurement		Data Analysis and Probability			
Apply transformations and use symmetry to analyze mathematical situations	Use visualization, spatial reasoning, and geometric modeling to solve problems	Understand measurable attributes of objects and the units, systems, and processes of measurement	Apply appropriate techniques, tools, and formulas to determine measurements	Formulate questions that can be addressed with data and collect, organize, and display relevant data to answer them	Select and use appropriate statistical methods to analyze data	Develop and evaluate inferences and predictions that are based on data	Understand and apply basic concepts of probability
X				X	X	X	X
	X						
	X					X	X
X				X	X	X	X
				X	X	X	X
				X	X	X	X
				X	X	X	X
X		X				X	X
		X	X				
		X	X				
		X	X				
	X	X	X			X	
		X	X				
X	X	X	X				

NATIONAL COUNCIL OF TEACHERS OF MATHEMATICS MATH CONTENT STANDARDS

Grades 4–12	Problem Solving				Reasoning and Proof		
	Build new mathematical knowledge through problem solving	Solve problems that arise in mathematics and in other contexts	Apply and adapt a variety of appropriate strategies to solve problems	Monitor and reflect on the process of mathematical problem solving	Make and investigate mathematical conjectures	Develop and evaluate mathematical arguments and proofs	Select and use various types of reasoning and methods of proof
1. Sona					X		
2. Cuica							
3. Madagascar Solitaire					X		X
4. Quipus	X	X	X	X	X	X	X
5. Counting Like an Egyptian	X	X	X	X	X	X	X
6. Breaking the Mayan Code	X	X	X	X	X	X	X
7. The Mayan Calendar Round	X	X	X	X	X	X	X
8. Weaving Baskets	X	X			X	X	X
9. Dyeing							
10. Tea and Temperature							
11. Rain in a Bottle							
12. From Arrows to Rockets			X				
13. Building Terraces							
14. Mud Bricks		X	X				

Every activity in this book addresses the Behavioral Studies Standard related to different ways that groups of people function. In addition, "Cuica" addresses the Art Standard about understanding connections among art forms and with other disciplines, and Music Standards related to performance, and about understanding the relationships among music and cultures. "Weaving Baskets" addresses Art Standards about understanding connections among art forms and with other disciplines, and about understanding the relationships among visual arts and cultures.

The Science Content standards are derived from National Research Council, *National Science Education Standards*, Washington, D.C.: National Academy Press, 1996.

The Math Content standards are derived from National Council of Teachers of Mathematics, *Principles and Standards for School Mathematics*, Reston, Va.: The National Council of Teachers of Mathematics, Inc., 2000.

Other standards references are derived from John S. Kendall and Robert J. Marzano, *Content Knowledge: A Compendium of Standards and Benchmarks for K–12 Education*, 3rd ed., Alexandria, Va.: ASCD, 2000.

Communication		Connections		Representation		
Organize and consolidate mathematical thinking through communication	Communicate mathematical thinking coherently and clearly to peers, teachers, and others	Understand how mathematical ideas interconnect and build on one another to produce a coherent whole	Recognize and apply mathematics in contexts outside of mathematics	Create and use representations to organize, record, and communicate mathematical ideas	Select, apply, and translate among mathematical representations to solve problems	Use representations to model and interpret physical, social, and mathematical phenomena
X	X	X	X	X	X	
X	X					
X	X	X		X	X	X
X	X	X		X	X	X
X	X	X		X	X	X
X	X			X		X
			X			X
X	X					

About the Authors

Exploratorium Staff Contributors

After years of teaching physics in institutions ranging from Princeton University to the Catholic University in Rio de Janeiro, **Maurice Bazin** codirected the Exploratorium's Teacher Institute from 1990 to 1995. During that time, he was the principal investigator for the National Science Foundation (NSF) grant that was the basis for this book. He now lives in Brazil with his wife and two children, ages eleven and three. He is presently involved in several Indian tribes' indigenous math and science education.

Modesto Tamez spent eighteen years as a bilingual classroom teacher, working with grades K–12, with an emphasis in teaching science. Since 1992, he has been working with the Exploratorium and with San Francisco State University, helping teachers integrate hands-on science into their curricula. Modesto was also co-director of Mission Science Workshop, an NSF-supported program to help establish after-school science programs in California. He is currently coordinating a mentor program, placing experienced teachers in middle school and high school classrooms to help new science teachers. He also teaches an elementary science methods course in a nontraditional intern program at John Muir Elementary School, under the auspices of San Francisco State University.

Additional Contributors

Chris Andersen got the idea for the "Rain in a Bottle" activity when he spent time in the Kalahari Desert of Botswana as a Peace Corps volunteer. Currently living in Paraguay with his wife and children, he teaches middle school science at the American School of Asunción. Before moving to Paraguay, he was in San Francisco, teaching middle school science and participating in the Exploratorium's Teacher Institute and the Mission Science Workshop.

Curt Gabrielson says he learned as much about science on his family's hog farm in Missouri as he did getting a physics degree from MIT. He worked at the Exploratorium's Teacher Institute for two years, separated by two years of teaching and learning in China. Curt ran the Taller de Ciencia/Science Workshop in Watsonville, California, where the majority of the population is Mexican or Mexican-American. The Workshop is a free community space where kids and adults have the opportunity to experiment with all sorts of science and art projects. He is currently teaching science in East Timor.

Paulus Gerdes is a Mozambican mathematician and educator who has written several books on geometry, mathematics education, and the history of mathematics in Africa. His books available in the United States include *Women, Art and Geometry in Southern Africa* (1998), *Geometry from Africa: Mathematical and Educational Explorations* (1999), and the forthcoming *Drawings from Africa* and *Sona Geometry*. He currently heads a research center on culture, mathematics, and education, is advisor to the minister of education of Mozambique, and was recently elected treasurer of the African Mathematical Union (AMU). He served as the president of the Pedagogical University of Mozambique from 1989 to 1996.

Robert V. Lange has a doctorate in theoretical physics and is a faculty member of Brandeis University, where he teaches basic mathematics in a program serving students in transition. After serving as a visiting professor at the University of Dar es Salaam in 1986, he was an adviser in science education to the Ministry of Education in Zanzibar and has consulted on the enhancement of access to science and education for women and girls in Africa. Since 1997 he has been coordinator for science and environmental education for the Ecumenical Patriarchate to create projects for clergy and others in the Black Sea region.

Kim Schuck is a traditional Native American basket weaver and holds an MFA in Textiles. She has worked with children for the last sixteen years, teaching math, science, art, and Native American studies. Her artwork is shown nationally and internationally. A professor of fine arts at San Francisco State Uni-

versity, she is currently working with the Asawa Fund to develop an elementary math curriculum based on origami. She spends much of her time raising her three children, two boys and a girl, who help to keep her focused.

Image Credits

Chapter 1: Sona Illustrations by Stephanie Syjuco and Stacey Luce. Photo (page 3) by Mário Fontinha; photo (page 6) by Maurice Bazin; other photos by Amy Snyder. **Chapter 2: Cuica** Illustrations by Stephanie Syjuco and Stacey Luce. Photos by Amy Snyder. **Chapter 3: Madagascar Solitaire** Illustrations by Stephanie Syjuco. Photos by Robert V. Lange. **Chapter 4: Quipus** Illustrations by Stephanie Syjuco; illustrations (pages 35 and 41) reproduced with permission from Marcia Ascher and Robert Ascher's *Mathematics of the Incas: Code of the Quipu,* Ann Arbor: University of Michigan Press, 1981. Photo (page 37) by Marcia Ascher and Robert Ascher/Artifact from the collection of the Museo Nacional de Antropología y Arqueología, Lima, Peru; photo (page 38) by Amy Snyder. **Chapter 5: Counting Like an Egyptian** Illustrations by Stephanie Syjuco. Photo (page 47) from the Metropolitan Museum of Art, New York; photos (page 49) reproduced from J. J. Clère, KEMI, *Revue de Philologie et d'Archéologie Egyptienne et Copte,* Paris, 1949; photo (page 57) copyright © The British Museum. **Chapter 6: Breaking the Mayan Code** Photo (page 61) copyright © The British Museum. Image (page 63) reproduced with permission from J. Eric S. Thompson's *A Commentary on the Dresden Codex,* Philadelphia: American Philosophi-

cal Society, 1972; photo (page 66) by Diane Burk. **Chapter 7: The Mayan Calendar Round** Illustrations by Stephanie Syjuco; illustrations (pages 71 and 73) reproduced with permission from J. Eric. S. Thompson's *A Commentary on the Dresden Codex,* Philadelphia: American Philosophical Society, 1972. Photo (page 72) by Modesto Tamez; other photos by Amy Snyder. **Chapter 8: Weaving Baskets** Illustrations by Stephanie Syjuco. Photos (pages 83, 85, and 90) by Maurice Bazin; other photos by Amy Snyder. **Chapter 9: Dyeing** Photos by Lily Rodriguez. Pyrex is a registered brand name of the Corning Consumer Products Company. **Chapter 10: Tea and Temperature** Illustrations by Stephanie Syjuco; illustrations (pages 107 and 111) reproduced with permission from *All the Tea in China,* San Francisco: China Books, 1990. Photo (page 109) by Bruce Dale/NGS Image Collection; other photos by Amy Snyder. **Chapter 11: Rain in a Bottle** Illustrations by Stephanie Syjuco. Photos (pages 119 and 123) by M. Shostak/Anthro-Photo; photo (page 120) by I. DeVore/Anthro-Photo; photo (page 121) by M. Biesele/Anthro-Photo; photo (page 125) by Peter Essick/AURORA; photo (page 126) by Christopher C. Andersen. **Chapter 12: From Arrows to Rockets** Illustrations by Stephanie Syjuco. Photo (page 131)

by Hillel Burger, Peabody Museum, Harvard University; photos (pages 132 and 135) by Arthur Barr, from the Frank F. Latta Collection, courtesy of the National Park Service, Yosemite Museum, Yosemite National Park; photos (pages 133 and 134) by Maurice Bazin; photo (page 137) by Modesto Tamez. Estes Model Rocket is a registered trademark of the Estes Corporation. Post-It is a registered trademark of 3M. **Chapter 13: Building Terraces** Illustrations by Stephanie Syjuco. Photo (page 151, left) by Amy Snyder; other photos by Curt Gabrielson. **Chapter 14: Mud Bricks** Illustrations by Stephanie Syjuco. Photos (pages 157, 158, and 162) by Tamia Marg; photo (page 161) by Robert V. Lange; photo (page 164) by Amy Snyder.